GENDER DIFFERENCES IN SCIENCE CAREERS

THE ARNOLD AND CAROLINE ROSE BOOK SERIES
OF THE AMERICAN SOCIOLOGICAL ASSOCIATION

GENDER DIFFERENCES IN SCIENCE CAREERS

The Project Access Study

GERHARD SONNERT

with the assistance of

GERALD HOLTON

RUTGERS UNIVERSITY PRESS
New Brunswick, New Jersey

Library of Congress Cataloging-in-Publication Data

Sonnert, Gerhard, 1957–
    Gender differences in science careers : the Project Access Study /
Gerhard Sonnert with the assistance of Gerald Holton.
        p.   cm. — (Arnold and Caroline Rose book series of the
American Sociological Association)
    Includes bibliographical references and index.
    ISBN 0-8135-2174-2
    1. Science—Vocational guidance—United States—Sex differences.
I. Holton, Gerald James.   II. Title.   III. Series.
Q149.U5S56   1994
502.3′73—dc20   94-25219
                                                              CIP

British Cataloging-in-Publication information available

Published by Rutgers University Press, New Brunswick, New Jersey
Manufactured in the United States of America

CONTENTS

FIGURE AND TABLES

Figure

Tables

FOREWORD

RIGHTLY ENOUGH, GERALD HOLTON, the distinguished progenitor of the Harvard Project Access that has given rise to this imaginative and exacting study, provides a preface, jointly with Gerhard Sonnert, that is as clear and direct as the book itself. Among other things, it introduces us to the varied highlights of the book. All to the good, one would think. Surely so for what I trust will be its many readers. But not for the fellow who had agreed to write a foreword; he is thereby condemned to little more than mere redundancy. When I plaintively pointed this out to Professor Holton, he promptly declared: "Not at all. You could still say a few words—[it was plain to me that he meant a very few words]—about the ways in which the study has drawn upon some of the ideas and findings in the sociology of science set forth by your colleagues, Harriet Zuckerman, Stephen Cole, Jonathan Cole, and yourself."

I take that as a mandate. Both for brevity and for limiting those few words to the many affinities of this work to the Columbia tradition in the sociology of science. All the more since this disciplined study of the centuries-old but only lately examined subject of gender differences in science careers also contributes to our understanding of the social structure of science and its consequences both for scientists and for scientific knowledge. And that is no small accomplishment. For as I have argued over the decades, the cognitive work of science takes place within a slowly changing institutional and structural framework. We must therefore understand such societal aspects of science as its evolving norms, sanctions, stratification, processes of competition, and reward-system if we are to acquire a fuller understanding of such cognitive aspects of science as choices of problems and concepts.

Happily, Holton's own method of "thematic analysis" can serve me well for identifying some sociological themes in this work that are not instantly evident. For, as Holton has observed, "Not all themata appear in so many words."

Thus, the theme of a "strategic research site" (SRS) appears tacitly in the Sonnert-Holton choice of a special sample of young women and men scientists for study: they are all recipients of prestigious postdoctoral fellowships. Insofar as this selective sample makes for similar qualifications and starting points, it provides an SRS for the investigation of gender differences in scientific career paths—that is, it provides fruitful data for investigation of previously stubborn problems and for the discovery of new problems begging for further inquiry. As a case in point, this sample of institutionally identified scientists of distinct promise provides an SRS for a deepened understanding of possible "glass ceilings" and "thresholds" for women scientists.

The fine-grained structural analysis adopted in this study greatly illuminates the theme of gender differentials in access to the opportunity structure of science. For a prime example, taking a postdoctoral fellowship in order to be with one's spouse puts both women and men in this strategic sample at a disadvantage in their later academic rank. Gerhard Sonnert then locates this gender-neutral finding in a further socio-structural context: many more women than men scientists are married to another scientist. This joint analysis results in a gender-specific finding: women scientists more often confront the "two-body problem" that limits their job choices geographically, a liability reinforced by the lingering cultural emphasis on according priority to husbands' career opportunities.

This book explicitly adopts the evolving sociological theme of the accumulation of advantages and disadvantages to integrate many of the findings on gender differences in scientific careers. That theme holds that initial individual and socially defined differences within would-be meritocratic systems of self and social selection accumulate to produce successively greater differences in access to the opportunity structure. In the domain of science, as in other institutional domains, initially small differences amplify in later stages of the individual career and aggregate into strongly skewed distributions of resources, role performance, and rewards. Within this theoretical context, as the authors point out, Jonathan Cole and Burton Singer have recently advanced a "kick-reaction model" to account for observed differentials in scientific productivity. A scientific career path is conceived of as an ongoing sequence of environmental events that provide individual "reactions" to those "kicks."

A great array of empirical findings in this work bear upon the gender-neutral and gender-specific advantages and disadvantages making for the differential accumulation of "human capital" in science. But it is not for me to deprive readers of the informed pleasure of coming upon these nicely differentiated findings and interpretations for themselves. Bon appetit.

ROBERT K. MERTON

PREFACE

THIS MONOGRAPH IS ONE of the chief products of Project Access, a large-scale research project based at Harvard University. The project was organized to investigate the detailed career paths of women and men in the United States who, after a promising start, headed for research careers in science, mathematics, and engineering. The project's major aims have been four: to compare women's and men's career trajectories; identify the obstacles and other reasons for comparative successes and failures; develop a theoretical framework for explaining gender disparities in science careers; and, from this research, to draw suggestions for policies that may help alleviate the inequities.

Project Access accumulated and used a data base—the largest of its kind—that was developed from two efforts. One was the solicitation of more than eight hundred responses to lengthy questionnaires (originally developed with the advice of Daniel Yankelovich and Dean Whitla and then field-tested). The other source was the result of open-ended, two- to three-hour individual interviews at their home sites all over the U.S. with two hundred men and women who had been matched as much as possible for such characteristics as field, academic age, and current occupation. These data were used to carry out the extensive set of statistical analyses concerning differences in men's and women's experiences in science that are reported in this book.

By studying the career trajectories of many individuals of both sexes, we were able to correlate career outcomes—for better or worse—with individual characteristics that might be predictors of the outcomes. Although we are confident that our work will also be of use to male scientists, we focused especially on those characteristics and events that seem to have a positive or negative effect on women's careers. We knew from the start that although institutional discrimination has been outlawed in the United States since the 1970s, impediments for women scientists have continued to exist—but they have become increasingly subtle.

This change in the structure of gender disparities in science careers suggested to us a modification of the more usual approach of studying them. First, when selecting a population for Project Access, we thought a study of a cross-section of the vast population of actual and potential scientists would be too coarse-grained and superficial to allow us to discern increasingly important subtle differences in career decisions and paths. We therefore focused on a relatively homogeneous subgroup of scientists who started their careers on a more or less equal footing. We chose to concentrate on the men and women who had been identified early in their careers as promising scientists, mathematicians, and engineers by virtue of having received one of the coveted postdoctoral fellowships of the National Science Foundation, the National Research Council, or the Bunting Institute at Radcliffe College.

All our respondents had demonstrated successful career paths through the first stages of a science career (up to the doctorate—a landmark at which many other hopefuls had already fallen by the wayside). But after their postdoctoral fellowships, career paths fanned out in different directions, some respondents becoming extraordinarily successful scientists, others less so, still others leaving science altogether. The career obstacles that confronted all these initially very promising male and female respondents—those that were eventually overcome as well as those that proved insurmountable—should be that much more worrisome and threatening to the many scientists who do not have the initial advantage of a prestigious fellowship. Thus, our respondents' career trajectories should contain useful lessons for anyone interested in either pursuing a science career or in administering programs that involve policy decisions about careers.

Second, the increasing subtlety of gender disparities in science careers also suggested a sort of triangulation from different methodological vantage points. The focus population was large enough for us to apply standard quantitative survey methods but small enough to let us study a sizable proportion in a more qualitative, in-depth fashion—that is, through open-ended interviews. (Our dissatisfaction with the limitations of purely quantitative methods is shared by other researchers in the area.) Adopting this unusual and burdensome double-barreled research model allowed us, over a period of several years, to gather and analyze statistically our questionnaire data. And we were also able to seek, through the interviews, the particularistic insights that only personal life histories can offer. In other words, we have supplemented the quantitative results with qualitative ones.

In addition to this monograph, we are publishing another book, *Who Succeeds in Science? The Gender Dimension* (Rutgers University Press, expected 1995). It has a different purpose from this book and is targeted to a wider audience, one consisting chiefly of those who are now in a science research career or aspiring to it. (A final section contains policy suggestions for improving the present imbalance of outcomes.) The book is intended to be not so much a contribution to sociological literature as a "how-to" book, a source of ideas and insights that may be of use to science-directed individuals in a variety of situations. We included the main findings that appear in much more detail in this monograph but also extensive, verbatim selections from twenty of the two hundred face-to-face conversations that our interviewer conducted with the respondents. We selected those cases for contrast. One group of ten consists of persons who had achieved successful careers as academic research scientists; the other ten had abandoned research science, even though their prestigious postdoctoral fellowships had initially marked them for probable success in science. Each group of ten includes five women and five men. The stories of their career paths, often told in the respondents' own pungent words, allow one to compare the strategies and experiences of men and women in science. They let the reader contrast those who achieved academic success with those who left research science. Of course, they also permit a glimpse into the dynamics and details of a great variety of real scientists' lives. We believe that these personal stories, all too rarely encountered in the literature, will have a special meaning and usefulness for the aspiring scientist.

Our study of the actual lives of contemporary scientists—presented here and in the other book—suggests that there are indeed strategies and characteristics that tend to favor the advancement of typical individuals and impede others. We also show that some "obvious" or "reasonable" decisions may have counterintuitive effects on subsequent careers. The reader will find that these and a considerable number of other, sometimes surprising, results clarify the forces involved in shaping the diverse career paths of scientists.

Our project was launched by a series of sociological and epistemological questions. Because the number of tenured women in physical science departments in the United States has been so small, we wanted to understand better the dynamics of careers in their early phases, which may handicap some scientists more than others who are equally gifted. In addition to the more obvious obstacles, were there differences in styles of approaching or doing research? Could one somehow tease out the various, rather small individual experiences that might lead in aggregate toward or away from the most desired career objective? (This corresponds to the highly suggestive notion of the accumulation of advantages and disadvantages first proposed by Robert K. Merton and Harriet Zuckerman.) Do the various theories of feminist critics of science find resonance among talented women scientists or explain their experiences to themselves in a way they regard as useful? The study of the history of science has taught us a good deal about how women fared professionally in the past, but in what ways would we find the current situation qualitatively (as well as quantitatively) different? Can a theory be developed that explains the complex interactions between structural obstacles and inherent "difference"? We would be gratified if our study helped to make these and a host of other questions easier to answer.

We take great pleasure in acknowledging the contributions of many people without whose work and help this project would not have been possible. Sara Laschever conducted the interviews throughout the country, and Mabel Lam developed the coding scheme for the interviews. Joan Laws ensured that the project would be kept on track administratively. Among the other fine members of our project staff, special thanks go to Robert Stowe, Sarah Tasker, and Sally Thurston.

Several members of our informal advisory committee—Jill K. Conway, John E. Dowling, Carola Eisenberg, Nathan Glazer, Kenneth Hoffman, Matina Horner, Lilli S. Hornig, Ellen J. Langer, Margaret L. A. MacVicar, Mary Bunting Smith, Shirley M. Malcom, Elizabeth McKinsey, Benson R. Snyder, Betty M. Vetter, and Dean K. Whitla—were kind enough to suggest improvements at early stages of the work, as were the directors of the Bunting Institute. We have already mentioned the help given in the design and testing of our questionnaires. We would like to thank J. Scott Long for his helpful comments on an earlier draft of our manuscript. Judith R. Blau, the editor of the Rose Book Series, and the other individuals involved in the Rose review process also have our gratitude. We gratefully acknowledge the financial support of the National Science Foundation, the Office of Naval Research, and the Ford Foundation. Of course, in no case did the advice or the grants imply an endorsement of our results. Last but not least, we thank the hundreds of women and men who generously gave their time in responding frankly to our questionnaires and interview requests.

The topic we have been studying has by no means been exhausted, nor has the potential for further use, by us or other qualified researchers, of our data and the full interview transcripts. All of these—with full protection of the privacy of the individuals—have been deposited at the Henry A. Murray Research Center of Radcliffe College.

Gerhard Sonnert
Gerald Holton
Jefferson Physical Laboratory, Harvard University

1

INTRODUCTION

AS PART OF HIS sociological classic *Soziologie. Untersuchungen über die Formen der Vergesellschaftung,* first published in 1908, Georg Simmel (1950) outlined a sociology of the *stranger.* His inquiry was abstract and general. Rather than describing specific strangers, it analyzed the sociological features of strangers within any social group. The very generality of Simmel's approach makes it a fruitful starting point for our investigation of women in science (notwithstanding Simmel's language, which now would perhaps be regarded as sexist). The key to understanding the persistent problems women face in science may be the extent to which they still are strangers in a strange environment.

According to Simmel, the stranger's position in a social group "is determined, essentially, by the fact that he has not belonged to it from the beginning, that he imports qualities into it, which do not and cannot stem from the group itself"(402). A crucial aspect of the stranger is "the more *abstract nature* of the relation to him. That is, with the stranger one has only certain *more general* qualities in common, whereas the relation to more organically connected persons is based on the commonness of specific differences from merely general features"(405).

Not only are the communalities between stranger and host abstract, but the stranger's differences from the host are also viewed abstractly. This makes for a facile transition from individual strangers to groups of strangers in the hosts' minds. Strangers are not perceived in their individuality but as members of groups. "The

1

consciousness that only the quite general is common, stresses that which is not common. . . . [T]his non-common element is once more nothing individual, but merely the strangeness of origin, which is or could be common to many strangers. For this reason, strangers are not really conceived as individuals, but as strangers of a particular type" (407). This perception is one of the key implications of what Kanter (1977a, b) described as the "token" status of minority members in organizations and professions.

In Simmel's essay, the stranger is "by nature no 'owner of soil'—soil not only in the physical, but also in the figurative sense of a life-substance which is fixed, if not in a point of space, at least in an ideal point of the social environment. Although in more intimate relations, he may develop all kinds of charm and significance, as long as he is considered a stranger in the eyes of the other, he is not an 'owner of soil'" (403). This concept goes far beyond the legal aspect of ownership. It also conveys an intangible bond—the feeling of being in a rightful place, even of belonging to this place.

For our purposes, Simmel's "owner of soil" translates into "owner of science." Thus, the central question of our project is, Have women become, alongside men, the owners of science? Of course, there are no longer any legal ownership restrictions for women to pursue science careers. But in terms of actual ownership, to what degree do women scientists' typical career paths still differ from those of men? And in the extended sense of ownership, we ask, Do women already own science and feel "at home" there, or are they still strangers in an area dominated by men? In other words, our study sets out to explore the extent of women scientists' *social marginalization* as well as its underlying causes.

Science, Gender, and Sociology

The subject of women's careers in science is located at the intersection of two fields of sociological study. The first is the sociology of science. In this framework, gender appears as one of the factors that determine scientific careers or activity. The second is the sociology of gender (often labeled gender studies or women's studies). In that framework, science appears as one of the societal arenas in which a gender division of labor exists or as one of the human activities influenced by gender. We hope that our data are useful to both sociological fields or any of their subunits.

To set the stage for our study, we now outline two major orientations in both of these fields. Sociological theories tend to be complex and idiosyncratic. A popular strategy for handling the complexity of sociological debates is to reduce them to dichotomies by marshaling varied theories into two opposing camps—with the obvious simplifications and distortions. Nonetheless, such a reduction is useful in providing rough reference points for individual studies. In this spirit, readers should understand the following outline not as a detailed review of individual theories, but as a simplified depiction of today's two principal sociological orientations. This depiction will help us locate our own theoretical approach and constitute a frame for using the data from the present study.

Both the sociology of science and the sociology of gender have in recent decades experienced considerable transformations. Their developments have strongly paralleled one another—an expression perhaps of the overarching, diffuse Zeitgeist that has shaped American sociology in general. In a gross simplification, we suggest a dichotomy between two orientations in sociology. The first orientation—best typified by Talcott Parsons's structural functionalism—is characterized by an interest in large-scale (macrosociological) problems, quantitative methodology, and objectivist epistemology. The second orientation, sometimes labeled postmodernist, consists of various strands predisposed toward small-scale problems, qualitative methodology, and relativist epistemology.

Sociology of Science

One can trace these two orientations in both the sociology of science and the sociology of gender. In the sociology of science, the first orientation is represented by the Mertonian research program initiated by Robert K. Merton, the founder of this sociological discipline. A major building block is Merton's identification of four core norms of science, the ethos of science—universalism, communism (of making scientific findings public; also called communalism), disinterestedness, and organized skepticism. In its emphasis on the normative structure of science, Merton's approach is somewhat related to Parsons's brand of structural-functional sociology. Issues of the social system of science (such as the differentiation of a social hierarchy among scientists) are the focus of this sociology of science; and on the methodological level, standard quantitative methods as well as qualitative methods are deemed appropriate to investigate them. In fact, the emphasis among the Mertonian sociologists of science is often on quantitative research, although a wide range of methods have been used.

On the epistemological level, cognitive and social aspects of science are demarcated according to a nonrelativist concept of scientific knowledge. For Merton, the institutional goal of science is the extension of certified knowledge—knowledge consisting of "empirically confirmed and logically consistent statements of regularities" (1973, 270). Such knowledge is essentially valid irrespective of the social aspects that led to its generation. This notion then allows the sociology of science to treat certified scientific knowledge at face value as a key input for the stratification in science as a social system. And by applying this notion of scientific knowledge to itself, the sociology of science claims solid ground for its own findings. Just as it is the business of the various scientific disciplines to generate certified knowledge about reality in their respective fields, the sociology of science aims to produce certified knowledge about the social system of science.

A main focus of the Mertonian research program has been to investigate to what extent the actual social system of science conforms to the basic norms of science. The norm of universalism in particular has received a great deal of attention. One of its tenets says that rewards within the social system of science (positions, prestige, prizes, and so on) should be allocated according to universalist—that is, meritocratic—criteria; and researchers have examined to what extent particularistic criteria come into play in the social reality of science. These researchers have been interested in the actual stratifying mechanisms that create a social hierarchy in science—in other words, in the causes explaining why some scientists become more successful than others. The Mertonian program has emphasized sociological mechanisms as opposed to psychological mechanisms.

Rather than referring primarily to scientists' personal attributes or traits, such as intelligence and creativity, or to the psychological mechanism of reinforcement as the source of social differentiation among scientists, the Mertonian program focuses on the properties of the social system of science.

A crucial sociological mechanism within the science system—and a major concept of the Mertonian research program—is the *accumulation of advantages and disadvantages* during the course of a scientist's career (Zuckerman 1989). This concept emphasizes feedback loops in real-life scientific careers. Small differences in earlier career stages are thought to amplify in subsequent stages and eventually lead to very different career outcomes. Two other Mertonian concepts can be considered special varieties of the overall process of the accumulation of advantages and disadvantages. The *Matthew effect* describes the propensity for already eminent scientists to receive disproportionately great credit for their scientific contributions while obscure scientists receive disproportionately little. For instance, if several scientists have rival claims to a scientific discovery, the most prominent among them is most likely to gain general recognition. Merton's *self-fulfilling prophecy*—a concept that applies to a variety of social situations and arenas—says in the case of science that young scientists may have a successful career simply because influential members of the scientific community have extolled their scientific potential. By helping the young scientists secure prestigious positions or research grants, the senior scientists' prophecy of scientific excellence may become true, even if it was not merited in the beginning.

Somewhat ironically, research into the stratifying mechanisms among scientists has complicated the goal of this line of research—determining the extent of universalism in science. The Mertonian research program has established the important role of past performance (track record and reputation) and of expected future performance (prophecy of scientific potential). Thus, universalism in science becomes hard to gauge because merit, the key ingredient in any universalist allocation of rewards, loses its quality as an independent, objective yardstick of scientific quality and becomes partly dependent on reputation and prophecy.

A much more radical questioning of universalism in science, as well as of other central tenets of the Mertonian program, has arisen from what is called the "constructivist sociology of science," "sociology of scientific practice," or "sociology of scientific knowledge" (see review by Zuckerman 1988). Of course, different scholars within the constructivist sociology of science differ markedly in approach and theory. The common denominator of the varied approaches under these headings is the tendency toward "postmodern attributions of subjectivity, relativism, and context-specificity" to science (Jasanoff 1992, 164). Those theorists hold that scientific knowledge itself is socially constructed, with only a tenuous (if any) relation to "objective" reality.

In contrast with the Mertonian interest in large-scale issues of the social system of science, constructivists put scientists "under the microscope." They typically study in exhaustive detail the actual behavior of scientists producing scientific knowledge. Main foci have been studies of laboratories (for example, Knorr-Cetina 1981; Latour and Woolgar 1979) and discourse and text analysis (Gilbert and Mulkay 1984). From that point of view, the Mertonian demarcation between social and cognitive aspects of science dissolves, and scientific knowledge is promoted from an external condition to a key issue of the sociology of science.

On the methodological level, constructivists often prefer qualitative over quantitative approaches. Assimilating anthropological and ethnographic ways, they generate descriptions of the minute social interactions and processes that lead to the generation of scientific knowledge. On the epistemological level, constructivists reject the notion of objective scientific knowledge and replace it with some sort of relativism. In this view, scientific knowledge bears the indelible marks of the social contingencies that shape its production; and sociological analysis reveals how scientific knowledge is socially constructed. The constructivist sociology of science thus fully appropriates the very core of science—knowledge production—and, in a reversal of the commonly perceived hierarchy of the sciences (from the "hard" natural sciences down to the "soft" social sciences) positions itself as a kind of metascience.

At this point, we should reiterate a word of caution: the dichotomy between the Mertonian program and the constructivist program has become a common reference point for sociologists of science. Conceptualizing theories as dichotomies carries the danger of over-polarization; and in this case, the differences between the two programs indeed may often have been exaggerated, as pointed out by Zuckerman (1988, 546–547).

Sociology of Gender

The sociology of gender intersects with both the sociology of occupations and, at a more general level, the sociology of social inequality. Division of labor is a key theoretical concept in these two sociological areas of research—who does what job in a society, and for what reason. The importance of a specific type of the division of labor, the division of labor by gender, has increasingly been recognized. Some sociologists studying inequality now conceptualize gender as one of the fundamental dimensions of social inequality (see Kreckel 1992, 1993); and attention to gender issues has similarly grown in the sociology of occupations. The gender division of labor is a central focus of the sociology of gender. From this perspective, our topic—women in science—can be seen as one instance of a much more general issue: women in professional and organizational hierarchies throughout society.

The dichotomy between the Mertonian and constructivist sociologies of science has its counterpart within the sociology of gender. Eisenstein (1983), for instance, has distinguished two waves in feminist theory. The first wave—paralleling the Mertonian sociology of science—has focused on the structural discrimination of women in various societal fields. Typical research topics include large-scale documentations of the extent of women's underrepresentation and discrimination in a variety of occupations. The methodology is predominantly quantitative, and the underlying belief in the objectivity of standard scientific knowledge remains unquestioned. An extensive body of research has found that across professions, women are overrepresented in the less prestigious ones and that within a given profession women tend to concentrate in the bottom ranks (see, for example, Bielby and Baron 1986; Reskin 1994; Roos 1985; Wright and Jacobs 1994).

Within first-wave analysis, attention has focused on the gender proportion in existing societal institutions—not on any deep- rooted gender traits. The similarity rather than the difference between the genders has been emphasized, opposing the conservative notion that women are "by nature" unfit for certain professional positions. The chief cause for women's underrepresentation has been identified in structural obstacles, and women have been deemed to be essentially equal to men so that a gender balance would develop within social institutions once the structural barriers were removed.

Whereas first-wave analysis examines the gender division of labor within existing institutions, the second wave—parallel to the constructivist sociology of science—focuses on women's distinctive traits, typically with a positive evaluation of them. People's patterns of behavior and thinking are considered to be thoroughly influenced by their gender—be it through biological differences or deep-rooted psychological or sociological processes. And major societal institutions are thought to carry typically male patterns at their core. Whereas first-wave analysis commonly argues that universalist (gender-neutral) institutional principles are corrupted by the institutional reality of gender discrimination, second-wave analysis regards these very principles as male.

We can now sketch how the issue of gender in science is framed in the traditional way (according to the Mertonian sociology of science and the first wave of feminism) and in the nontraditional way (according to the constructivist sociology of science and the second wave of feminism). Traditionally, sociologists have studied the gender division of labor in science in terms of disparities between women scientists' and men scientists' career outcomes. Within the framework of Merton's basic norms of science, such a division of labor based on gender is conceptualized and examined as a *deviance* from the norm of universalism. In other words, science is intrinsically gender blind, but structural barriers against women scientists have introduced the pathology of gender disparity into the social system of science. The Mertonian research program has spawned a sizable research literature that has examined gender disparities in the reward system of science—that is, to what extent women have been disadvantaged in positions, salaries, other rewards, and so on. Typically, these studies use a large-scale statistical approach to document differences between the genders at various stages of science careers.

The constructivist sociology of science and the second wave of feminism converge in giving gender a much more central and far-reaching role in science. Gender is seen to pervade its institutional and even cognitive aspects. The institutions of science are considered deeply shaped by the fact that men have created them and predominated in them. And on the epistemological level, gender is said to be one of the factors that leaves its mark in the very construction of scientific knowledge; hence, there are male and female ways of knowing. A science controlled by women might construct knowledge in different ways, embodying a different epistemology. From a constructivist perspective, Merton's basic norms of science are viewed as an expression of a particular male-oriented science. The norms of disinterestedness and organized skepticism, for instance, may be considered rooted in a specifically male epistemological approach.

Gender and Stratification in Science: Deficit Model versus Difference Model

Having outlined what we consider the major fault lines in the general territory of the sociology of science and of the sociology of gender, we now expand on models explaining why women scientists as a group have had less successful careers than men scientists have. The various models that have been developed to address this question can be categorized into two basic types: the deficit model posits that women are treated differently in science, and the difference model says that women act differently in science. In Simmel's terminology, the deficit model describes how women are treated as strangers in science, and the difference model shows how they act as strangers. The two models will serve as the theoretical reference points for our analysis of the careers of the female and male scientists in our study.

Because of the different ways in which the gender issue is conceptualized, the traditional orientation typically offers explanations related to the deficit model, although it sometimes also includes elements of the difference model. The nontraditional orientation has a greater affinity to the difference model. We shall now see how these two sets of theoretical concepts play out in a review of existing research that provides the background for this study.

Deficit Model

The deficit model is based on structural explanations of scientific careers. It focuses on formal and informal exclusions of women scientists. According to this model, women as a group receive fewer chances and opportunities in their careers; and for this reason they collectively have worse career outcomes. The emphasis is on structural obstacles—legal, political, and social—that exist (or existed earlier) within the social system of science. It is assumed that women's goals are similar to men's goals, but that barriers to advancement keep women from accomplishing these goals on par with men.

Within the deficit model, structural obstacles can be ordered along a formal-informal spectrum. Formal structural obstacles are the most powerful ones in keeping women from career success. In the past, for example, the admission rules of many institutions of higher learning denied women access to education and thus severely restricted their career opportunities in the sciences. Open and outright gender discrimination, which denies women good entry-level jobs, promotions, tenure, and research funding on account of their gender, can also be considered a formal structural obstacle. While these obstacles are the hardest to overcome for individual women, they are also the easiest to be identified—and to be removed, if there is a will to do so.

During the past two decades, formal structural barriers have been outlawed in the United States, and gender disparities in the sciences have subsequently decreased. Nevertheless, full gender equity is elusive. As a result, attention has recently turned to the more subtle informal barriers that women face within the social system of science. Published research findings generally agree that women scientists are still handicapped by these informal structural obstacles. For example, women have less access to strategic resources, such as social networks, that are essential for career success. Compared with men scientists, women scientists on average may be more socially isolated from (usually male) mentors during their training and from the network of colleagues during later phases (Epstein 1970; Kaufman 1978). Thus, women tend to remain in the "outer circle" (Zuckerman, Cole, and Bruer 1991)—outside of the influential clique of scientific powerbrokers, the key researchers and administrators who make important decisions about the future of a research field or academic discipline. Even in the absence of any intentional exclusion, women's minority position in a male-dominated field may carry with it the adverse effects of tokenism (Kanter 1977a, b). Thus, although outright exclusion has subsided, inequities—sometimes called micro-inequities (CSWP 1992b)—may still persist and result in differential career outcomes. Cole (1987, 368) noted that it is "in the domain of informal activities in science that the biggest gaps between men and women remain. It is in the more intangible set of experiences associated with doing science from day to day that women rightly feel most excluded." In Simmel's terms, women still tend to be treated as strangers in science. A crucial theoretical concept is that of the critical mass. When strangers become more numerous, they may eventually reach a critical mass beyond which they are strangers no longer. In science this means that when a sufficient number of women scientists has been reached in a department or a field, they lose their status as tokens or oddities. Dresselhaus (1986), for instance, observed that isolated women in physics classes used to be very taciturn; but once the proportion of women reached a critical mass of around 10 to 15 percent of the class, their level of participation became indistinguishable from that of the men. If women scientists become an integral part of a mixed-gender environment, their presence will affect the social conditions of science in various ways, from the style of day-to-day interactions to implementing policies on day care and to tenure decisions.

Finally, whereas formal and informal structural barriers may directly affect the careers of a number of women scientists, they may also indirectly affect an even larger number of women by turning them away from a career in science. If women perceive that because of structural obstacles in the social system of science the potential rewards of a scientific career are scant or uncertain, it may be a rational choice for them to avoid a career whose potential costs appear to outweigh its potential benefits (Moen 1988; Weitzman 1984).

Difference Model

We now turn to the difference model, which emphasizes deeply ingrained differences in behavior, outlook, and goals of women and men. According to this model, the root cause of gender disparities in career achievement is internal to the individual. It lies in gender differences—either innate or the result of gender-role socialization or cultural patterns—which thoroughly shape individuals' behavior as well as the character of social institutions. In this study we concentrate on socialized or cultural gender differences.[1] Among the variety of such internal differences mentioned in the literature, three categories can be distinguished. First, females may be more likely than males to be socialized with general orientations and attitudes that reduce the drive toward professional success in any field. Second, particular attitudes about science may define it as a male field and thus encourage males and discourage females to participate. Third, deep-seated epistemological gender differences may make science as it is practiced today insufficiently compatible with women's ways of thinking. Let us examine these points briefly.

Traditional gender-specific socialization is based on, and reinforces, the fundamental division of labor by gender: women have primary responsibility for the home and family, and men are breadwinners. This traditional division has been considerably weakened in recent times, particularly through the increasing participation of women in the labor force. In spite of emerging egalitarian patterns, however, it has not completely disappeared. Even though more women have become primary wage earners, gender differences in career patterns remain. The normative interpretation of these persisting differences posits that with the weakening of traditional role divisions women now experience a broader range of culturally acceptable options. Whereas men's main avenue to respectability and success remains their career, for women both career and homemaking represent acceptable pursuits (Cole and Fiorentine 1991). In a more structural view, women's wider range of acceptable options is interpreted as a wider range of responsibilities. In the majority of dual-career couples studied by Hochschild (1989), it was the women who had to fulfill most of the domestic duties.

Even with the decline of rigid gender-role socialization, social practices still reinforce the image of the aggressive and successful man and the nurturing and supportive woman. From early on, for example, girls tend to be discouraged from developing a strong motivation for achievement (Eccles-Parsons, Adler, and Kaczala 1982; Entwisle and Hayduk 1988; Lipman-Blumen 1972). One factor in reducing women's career ambitions may be the influence of protective parents and teachers who have lower expectations for females (McBay 1987). Compared with males, most females appear to form different values and goals that shape their notions of success and their career and life decisions (Eccles 1987). As a group, women may have a lower level of career aspirations, but this is not to say that ambition is alien to them. Rather, women's ambitions may focus more on areas traditionally defined as feminine, or their ambitions may be vicarious (Keohane 1984; Stein and Bailey 1973). Examples of the latter case are women who project their ambitions onto other members of their family—husband or children, especially boys. "Since a woman often puts aside any personal ambitions for motherhood, the child may be expected to succeed in her stead, to act out her ambitions for her" (Hoffnung 1984, 130).

Psychological research has found that individuals' attributions or beliefs about the causes of success and failure have major effects on their achievement-oriented behavior (Frieze 1978; Weiner 1974). Those who attribute their success to internal factors (their own high ability and effort) and failure to external factors (bad luck and task difficulty) are more likely to be achievement-oriented than those with the reverse attribution pattern. The genders appear to differ in their attributions. Compared with males, females tend to emphasize luck in the explanation of success and low ability in the explanation of failure—a less beneficial attribution pattern that was empirically found in some, albeit not all, studies of the issue (Frieze 1978; Frieze, Fisher, Hanusa, McHugh, and Valle 1978). Similar attribution differences exist in the evaluation of others: a well-performing girl is perceived as lucky, a well-performing boy as gifted. Thus, females are socialized to be less resilient and more easily discouraged than males. Widnall's (1988) observation that female graduate students tend to have feelings of frustration and discouragement when facing adversity, whereas men are more prone to feelings of anger and rage, may be an effect of attitudes like these.

In a related result, Salk (1989) found that among students at an elite business school, those from a high socioeconomic background tended to attribute their achievement to their own prowess, whereas those from a low socioeconomic background tended to attribute their achievement to luck. It appears that not only women but, more generally, members of social groups who are not socially or culturally *expected* to succeed are prone to equate success with luck; people who are already privileged in some way seem to assume that success is part of their birthright.

From early on, girls are socialized to interact in a style that de-emphasizes aggressiveness and competitiveness, whereas those characteristics are encouraged in boys and become embedded in a male interaction style (Tannen 1990). Chodorow (1974) traced gender differences—women's greater connectedness with others and men's greater independence—back to early childhood dynamics. In this view, processes of individuation and separation from the mother are more central for young boys (who need to form a different, male, identity) than for young girls. These socialization patterns tend to distance women from precisely the characteristics—such as ambition, self-confidence, resilience, aggressiveness, and competitiveness—that the current social system of science rewards and reinforces.

In addition to socialization practices that may impede women in achieving success or deter them from its pursuit in any professional field, specific cultural beliefs about science and scientists may distance women from this field in particular. Science is commonly perceived as a thoroughly male domain. Scientific textbooks have often reinforced this notion (at least until recently) by mentioning and picturing almost exclusively males and by showing the few females who do appear in gender-stereotypical roles (Heikkinen 1978; Kelly 1985). As a result, female students lack role models of successful women scientists and may be discouraged from pursuing scientific interests.

Furthermore, juxtaposed to the positive image of science in textbooks is a powerful antiscience theme in popular culture. A common stereotype of the scientist is the "egghead" or, more fashionably, the "nerd"—an undersocialized and barely functional person whose scientific pursuits leave him grotesquely detached from the real world. However contrary in other respects, this negative image shares with its positive counterpart the characterization of the scientist as male. As Brush (1991, 406) concludes, it is "difficult to imagine that the incessant pairing of 'scientist' and 'man' has no effect on young women." A young woman with scientific aspirations may face a double marginalization: entering the stigmatized subculture of nerds, and then being an oddity even among her fellow nerds because of her gender.

Some scholars see the maleness of contemporary science expressed not only through who the scientists are and how they socially interact, but even through how they do science. According to this view, the cognitive domain of science is influenced by the scientist's gender, and current scientific epistemology and methodology are androcentric. Science is said, for instance, to embody a masculine type of objectivity and rationality; women's epistemological style would be different, more intuitive, synthetic, and holistic (see Belenky, Clinchy, Goldberger, and Tarule 1986; Keller 1983, 1985, 1989; Kerr 1988). If women have a different methodological approach to science and different ideas of what constitutes good science, the fundamental incompatibility of this view with the current, supposedly androcentric system might handicap them severely. Keller (1983) portrayed the biologist Barbara McClintock to be an example of a scientist with a particularly female style. But among the various types of women's obstacles mentioned, the hypothesis of an androcentric epistemology in current science is one of the most controversial and has encountered vigorous opposition (for example, Levin 1988).

In closing this section about the difference model, we note Weitzman's (1984) warning against painting an oversocialized portrait of women as prisoners of socialization. Gender-role socialization does not completely determine the outlook of women. First, it is overlaid with a variety of other socialization patterns, especially in a country such as the United States, with its multiple ethnic, cultural, and religious traditions and class divisions. Moreover, some room is always left for individual differences, much more so when traditional gender roles are in flux. In particular, the women in our sample who successfully completed a doctorate in the sciences and won a competitive postdoctoral fellowship cannot be viewed as typical products of traditional gender-role socialization.

For reasons of conceptual clarity, we have distinguished between obstacles that exist as structural features of the social system of science and those that exist in the form of women's attitudes and behaviors. But in reality these two categories are intertwined and reinforce each other, which most scholars in this field acknowledge. The differences between these researchers' stances are less of principle than of degree—they differ in the relative weight they attach to structural and individual explanations.

On the one hand, the cultural attitudes and values of others (parents, teachers, and even peers) may translate into structural obstacles. For instance, teachers who are convinced that science is an unfeminine profession may treat girls as strangers who do not really belong. They may focus their attention on boys and tend to give rewards, such as prizes and scholarships, to boys rather than girls; their acts then become a structural obstacle for girls. If, on the other hand, girls who come up against a series of such structural obstacles finally get the message that they are not wanted in this area, they may internalize those structural barriers by revising their own outlook, their aspirations, and self-confidence.

Nevertheless, if we are to understand the complexities of women scientists' career paths, both models of career achievement are needed—the deficit model, with its focus on structural obstacles, and the difference model, with its focus on internal differences between men and women. Cole and Singer (1991) developed a highly formalized *kick-reaction* model of a scientific career path to account for scientific productivity. Although we do not wish to use this model in its mathematical rigor, the basic kick-reaction concept is useful in conceptualizing how scientific career paths in general are shaped. A career path can be seen as a sequence of kicks from the environment and reactions to those kicks by the individual. Among the kicks, we count any event in the environment that has an effect on the individual, be it a potential obstacle (or discouragement) or advantage. The deficit model focuses on the negative kicks women scientists receive, whereas the difference model focuses on women scientists' disadvantageous reactions.

In the following we use the dichotomy of the deficit model and the difference model in a brief chronological overview of previous research on women in science. This overview has two particular purposes. First, it illustrates the notion that the career path into the sciences is a dynamic, complex system in which gender disparities from a whole series of stages may accumulate. As a group, females with an interest in science face barriers throughout the various stages of the science pipeline. Second, it demonstrates that, at nearly every stage, structural and internal obstacles combine and reinforce each other.

A Chronology of Obstacles

Crucial socialization processes take place in the family environment, and a large body of research has pointed out that many gender differences (for instance, the previously mentioned gender differences in achievement orientation and self-confidence) may to a considerable extent be formed in the family when a child is young (see Huston-Stein and Higgins-Trenk 1978). School has also been identified as an important agency instilling cultural sex roles that distance girls from academic achievement in general and from achievement in the sciences in particular. Both teachers and peers may contribute to it (Weitzman 1984, 183–193). To some degree, there may also be lingering structural barriers that keep girls from receiving additional or specialized instruction in the sciences.

In college, women tend to experience a variety of pressures that work against developing or maintaining a career interest in the sciences. Structural obstacles, such as a hostile classroom atmosphere (Hall 1982), may combine with more internal concerns about not living up to the stereotypical image of the popular college woman, which does not normally include a serious involvement in the sciences. Graduate school is another important phase in a scientist's education. Here again, women may face structural obstacles, and science orientations may be weakened. At the structural level, insufficient financial support for women relative to men appears to be a key cause for women's lower participation in and lower completion rates of graduate studies (Hornig 1987; Widnall 1988). Furthermore, there may be a dearth of formal and informal support (Dresselhaus 1986; Kistiakowsky 1980; Kjerulff and Blood 1973; McBay 1987; Moen 1988; Widnall 1988). At the internal level, many women students in graduate school also find it difficult to cope with a milieu characterized by hostility and aggressiveness—the "combative style of interaction" (Widnall 1988, 1744; Hall 1982). The high degree of overt combativeness may be a corollary of the typically masculine interaction pattern mentioned earlier (Tannen 1990). Furthermore, women graduate students were reported to suffer from an "'imposter' syndrome" (Widnall 1988, 1743), feeling a deep sense of inadequacy even if it is objectively unfounded (Briscoe 1984, 154; Hornig 1987; Widnall 1988).

The professional career appears to be the phase during which some women scientists experience the clearest cases of structural obstacles. These obstacles come in the form of discrimination in hiring, promotion, granting of tenure, or funding of research. Whereas instances of open discrimination are rare (because it carries heavy legal penalties), subtle and informal structural barriers may linger, as mentioned earlier. In many fields, women scientists tend to be marginalized in both the formal and informal scientific communication systems (Moen 1988). They find it harder to collaborate with colleagues (Reskin 1978). They appear to lack access to resources, positions, and power (Briscoe 1984). They face prejudice from the "old boys" (Koshland 1988) and are excluded from the inner circle of the scientific establishment (Kistiakowsky 1980).

Family Issues, Private Life, and Career

The issue of marriage and parenthood is often considered one of the major obstacles for women scientists on both the structural and internal levels. The integration of career and family life is a highly complex issue. Different authors vary in the relative attention they give to structural or internal obstacles, but elements of both the deficit model and the difference model may combine here. Many young women consider marriage and motherhood desirable goals for which they are willing to make career sacrifices. On the other hand, they may be forced to play the roles of housewife and mother unassisted because of their husbands' career requirements, attitudes, or behaviors.

A predominately structural barrier is that women scientists interrupt their career more frequently than men scientists do (Lewis 1986), often because of the responsibilities of raising a family. Although pregnancy and childbirth are biological functions of womanhood, the fact that in our society women usually have the chief responsibility for raising children is not necessitated by biology but based on a deep-seated societal division of labor by gender. A widespread stereotype reported among male professors is that female students lack commitment and are prone to leave science in favor of raising a family (Widnall 1988). On the other hand, women scientists who do persist in their careers, even when they have a family, may appear to violate societal role expectations. They may be criticized for neglecting their children (Koshland 1988) or feel guilty about it themselves.

A much higher proportion of women scientists than men scientists are married to another scientist, very often in the same field. Hence, for a large number of women scientists the "two-body problem" of finding two science positions in the same geographical area is an important issue. In dual-career marriages, priority is typically given to the husband's career opportunities (Ferber and Huber 1979; NRC 1981; Rosenfeld 1984). Thus, one of the structural barriers imposed by marriage is that married women scientists are more geographically restricted in their job choices (Moen 1988; NRC 1981).

In the light of the preceding discussion, as well as intuitively, one might assume that marriage and child rearing hamper women's science careers. Nevertheless, research findings on this topic are mixed and inconclusive. (For a review of relevant research, see Long (1990, 1299.) We will examine the interplay of marital and parental status with the career path among both the women and men in our group.

Discrimination

Gender discrimination is one of the most evocative terms to explain gender disparities. Therefore, we need to clarify the notion of gender discrimination before presenting our results. In particular, the two fundamental orientations of the sociology of science (as described earlier) translate into drastically different notions of gender discrimination. Two basic notions can be distinguished: process theory and substantive theory (Cole and Fiorentine 1991; Fiss 1991). Whereas process theory focuses on the criteria in a selection process to determine whether it is discriminatory, the substantive theory focuses on the outcome of a selection. For a start, the first can be aligned with the Mertonian orientation, the second with the constructivist orientation. But we will see that things get rather more complicated in the course of our discussion.

At a sufficiently abstract level of definition, the two concepts of discrimination agree. Cole's (1979, 50) definition of discrimination, for instance, appears fairly noncontroversial: it is defined as "the importation into a social situation of characteristics that are functionally irrelevant." The two concepts part ways, however, over what is considered functionally irrelevant—or, more specifically, whether gender is functionally irrelevant to science.

Key determinants of functional relevance or irrelevance and, consequently, of the concept of discrimination in science careers are the assumptions about the nature of scientific knowledge. If scientific knowledge progresses toward a more thorough and complete understanding of the world (as the Mertonian approach presupposes), there is, at least in theory, a basis for an objective measure of scientific performance—a scientist's contribution to the advance of scientific knowledge. Objectively determined performance, then, is considered functionally relevant for the social stratification among scientists. By contrast, gender, along with race, nationality, religion, and similar ascriptive characteristics, is functionally irrelevant. Women who are excluded from scientific rewards (jobs, promotions, grants, and so on) because of their gender are considered to be discriminated against. But women who are excluded because of their *performance* as scientists are not considered to be discriminated against.

Just as an objective performance standard leads to the process theory of discrimination, its absence leads to the substantive theory of discrimination. Within a constructivist framework, objective knowledge is impossible. At most, there are contextual performance criteria that are valid only within a certain group or school of scientists and cannot claim validity across different schools. If objective performance criteria do not exist, anything could be functionally relevant for the construction of scientific knowledge. Gender, among a host of other criteria, is a prime candidate for being functionally relevant as a determinant of scientific knowledge. The allocation of rewards becomes a political struggle between different ways of constructing scientific knowledge. As a consequence, allocation differences themselves—that is, gender differences in career outcomes—become coterminous with gender discrimination. According to this view, women's numerical underrepresentation in science or in the upper echelons of science is already a sign of discrimination by a male science establishment with its arbitrary performance standards that favor men.

In the substantive theory of discrimination, the issue of functionally relevant performance criteria is decided as a matter of principle: objective performance criteria are utterly impossible. In the process theory of discrimination, the same issue is pragmatic—a measurement problem. But this problem turns out to be far from trivial. For instance, the number of publications may not accurately reflect contributions to knowledge. And there looms an even larger problem in the very structure of scientific careers. As we have already mentioned, the discovery of the complex feedback processes that shape science careers (captured in the notion of the accumulation of advantages and disadvantages) has made it much harder to assess performance or merit at any given time. Applied to the issue of gender discrimination, this means that prior discrimination may contaminate performance measures as well as other innocuous-looking criteria, such as skill and experience.

To illustrate this difficulty in the process model of discrimination, we draw on the distinction between *dual labor market* theories and *human capital* theories of explaining gender inequalities at work (Blau & Jusenius 1976). The first theory type is structure oriented and emphasizes discriminatory hiring and promotion practices; its notion of gender discrimination is fairly straightforward. The second theory type is oriented to the individual and emphasizes that women accumulate less human capital in terms of education, training, experience, and skills; this is where the notion of gender discrimination becomes difficult.

Like Cole (1979), many proponents of the process theory of discrimination tend to approach the issue of gender discrimination in science as "sophisticated residualists." They control career outcomes for performance as well as for various human-capital aspects, such as the caliber of graduate school, the prestige of mentors, early research experience as research assistant, and similar variables. Only if a gender disparity remains after these controls is it considered a discrimination.

As Cole (1979) himself pointed out, however, such sophisticated residualism quickly leads into murky areas. If scientific careers are indeed protracted and dynamic processes of accumulating advantages and disadvantages—human capital—then discrimination might be hidden in earlier career stages where women may have been denied equal access to human capital. Moreover, women may be less able to benefit from early advantages or more likely to be severely punished for early disadvantages in their later careers. In the language of capital, women may receive lower interest rates on their assets than their male counterparts do, or they may be charged more interest on debts. Thus, although human-capital criteria, such as skill and experience, appear to be universalist and pertinent on the surface, they might conceal prior discrimination. And this realization may make the outcome approach—the substantive theory—more attractive to some of those who do believe in the functional relevance of performance.

So far, we have dealt with a statistical approach to discrimination—the dependent variable being outcome disparities between groups of women and men scientists. But there is another approach to discrimination: asking women scientists themselves about their experience with discrimination. Of course, in some cases, women may not recognize when they are discriminated against; in other cases, they may see discrimination where there is none. This subjective approach becomes more important, however, as the link between discrimination and career outcomes becomes weaker. In other words, if women are as successful as men, this does not necessarily mean that they have not encountered any discrimination. They may have been able to counter it successfully.

Our Approach and Research Site

Having outlined the two major positions in the sociology of science and the sociology of gender, we can now locate our own approach in this framework. On the epistemological level, our stance is decidedly Mertonian—that is, nonrelativist. This is not to postulate pure knowledge in the sense of absolute, objective truth. We acknowledge multiple social influences on research and scientific knowledge but emphasize that scientific knowledge is greatly shaped by the condition of the world, which is essentially invariant with respect to transformations into the individual observers' reference frames. In comparison with various other types of knowledge in society such as political ideology and religious belief, we believe that the findings constituting established scientific knowledge are clearly and vastly more independent of the social factors conditioning these other types of knowledge.

Constructivists rightly point to the social component of every construction of reality; but they seem more interested in the iconoclastic demolition of the straw man of an absolute-truth claim of science than in acknowledging the high degree to which established scientific knowledge is related to data not dependent on social, political, philosophical, or other presuppositions. Furthermore, the relativist sociology of scientific knowledge faces the full brunt of what is known as the problem of reflexivity: if there is no objective scientific knowledge, how can the sociology of scientific knowledge itself— the very concept that there is no objective scientific knowledge— claim validity?

In terms of scale and methodology, we take a pragmatic and pluralist position. In our view, social reality itself suggests certain approaches. The scope of the research question and the methodology employed should change with the changing social conditions. For example, the situation of women in science has changed dramatically during the past two decades. Women scientists are much less often the victims of formal structural obstacles; the obstacles have become more subtle and minute. It makes less and sense to view women scientists as helpless individuals buffeted by adverse structural forces beyond their control. And the social system of science contains obstacles but increasingly also opportunities for women scientists. All these changes in the social reality suggest widening the scope to include both a macroanalysis of basic career outcomes of large groups of men and women scientists and a microanalysis of career paths and career decisions. On the methodological level, they suggest supplementing quantitative statistics with a more fine-grained approach that looks into the behaviors, attitudes, and norms of scientists in greater detail, also using qualitative methods and learning from individual accounts.

Now we explain why we selected former recipients of NSF and NRC postdoctoral fellowships as the population of our study. After discussing the theoretical issues that underlie this selection, we briefly address two points. One concerns the institution of the postdoctoral fellowship, which was established rather recently as a major career stage and which may fulfil various functions in a science career. The other point concerns placing our group of women scientists in the context of the sweeping changes in women's situation in the sciences.

Taking into account the complex process of the accumulation of advantages and disadvantages during science careers, the dearth of successful women scientists in many fields of science may have resulted from a multitude of reinforcing factors. To start disentangling this web of causality, we concentrate on only a segment of the career path, which is the portion after the end of formal training. We try to control for prior differences by examining the careers of men and women scientists who had similar qualifications and starting positions at the beginning of their professional careers.

More specifically, we concentrate on a very promising subpopulation of young men and women scientists in the U.S. with similarly auspicious preconditions for a science career. The extent of gender-specific divergences in the further career paths of scientists who succeeded in obtaining a prestigious national postdoctoral fellowship from the NSF or NRC is the topic of our study.

Why do we concentrate on the top end of the general scientist population—on a kind of elite sample? Elites hold, of course, a great deal of intrinsic interest for sociologists simply because elites tend to have a disproportionately large input into society. In this sense, our study previews some characteristics of the American scientific elite that will be prominent during the coming decades. But there are more specific theoretical reasons for studying such an elite sample— the descriptive concepts of *glass ceiling* and *threshold* that denote two distinct patterns of deviance from a pervasive gender-neutral accumulation of advantage and disadvantage.

As large numbers of women entered science careers during the past decades, interest increasingly focused on the end points of their careers. Glass ceiling describes an invisible but real barrier that is thought to impede women from reaching top positions in the professions. The concept is very general in scope, maintaining that the same mechanisms are at work not only in science but also in many other areas, notably business, where Kanter's (1977a) classic *Men and Women of the Corporation* addressed issues of this kind. Our study of the glass ceiling in science may therefore serve as a point of comparison for similar studies in other areas.

The glass-ceiling concept posits that, whereas women may be accepted at the lower and intermediate ranks of a hierarchy, it is much harder for them to enter top positions. Women scientists who have been awarded prestigious postdoctoral fellowships should have accumulated significant advantages up to that point and should be highly qualified, as well as motivated, to pursue a successful research career in the sciences. If there is a glass ceiling for women scientists, our sample is a prime research site to detect it. If these promising women scientists in our sample turn out to be less successful as a group than comparable men, they may have encountered a glass ceiling of gender-specific obstacles in the later stages of their professional careers. We should add, referring to our discussion of gender discrimination, that we will not automatically equate a glass ceiling of gender disparities in career outcomes with gender discrimination.

The alternative model to the glass-ceiling notion of career development is that of a threshold. According to this model, the processes of professional stratification would be gender neutral for those relatively few promising women who succeeded in overcoming certain earlier barriers—in this case, by having obtained a prestigious postdoctoral fellowship. In other words, gender differentials are thought to be particularly large in the lower, not the upper, ranges of achievement. According to this concept, one could surmise that once women have survived in the science pipeline to the stage where they received prestigious postdoctoral fellowships, they will, on average, not face gender-specific obstacles during the later stages of their careers. These women would be said to have passed a threshold beyond which gender no longer matters.

Whereas there is a large amount of prima facie evidence for the glass ceiling, some empirical observations from various areas lend plausibility to the threshold concept: "high-quality" women succeed to a similar degree as "high-quality" men, whereas gender becomes a handicap for less outstanding women. Cole (1979, 75–77) found that gender did not predict the prestige of institutional affiliation for a general sample of scientists but did so for scientists who had not published a single paper during the first seven or eight years of their careers. Nonpublishing women scientists worked in locations of lower prestige. Cole (1979, 75) hypothesized that in cases in which there is no strong evidence to judge a scientist according to meritocratic performance criteria, the functionally irrelevant gender status is activated. In a study of college admissions, no gender difference was found in admissions at the high-ability level, but males were preferred over females at the low-ability level (Walster, Cleary, and Clifford 1971).

Supportive evidence of a different kind appeared in a study of career-plan persistence of pre-med students (Cole and Fiorentine 1991). Among pre-med students with excellent grades, the proportion of applicants to medical school did not differ by gender; however, fewer women than men with marginal qualifications applied to medical school. In contrast to the college admission study, where gender differences resulted from admission committee decisions, Cole and Fiorentine (1991) explained the gender difference in career-plan persistence by a difference in norms. According to them, men are under greater pressure to succeed professionally because this is the only way they can be successful, whereas women have a wider range of normative alternatives. The hypothesis that women are influenced by the wider range of socially approved options was also discussed by Bernard (1964).

By concentrating on men and women recipients of prestigious postdoctoral fellowships, we control for high promise at this career stage. But this broad control does not, of course, guarantee that the groups of our women and men respondents are perfectly alike in achievement and promise. On the one hand, although the NSF and NRC fellowships are open equally to women and men, women may receive a marginal bonus in the selection process. For instance, in selecting fellows from the group of applicants who were judged to be borderline qualified, NSF considers a variety of factors, among them gender. This slight bias in favor of women—lowering the women awardees' average qualification—might be counteracted by a bias against women—raising the women awardees' average qualification—throughout the science pipeline. If the accumulation of advantages and disadvantages is stacked against women at many stages, then the surviving women may be of higher caliber than the men, who, on average, have traveled a less difficult path. Here we propose a compensation model (to which we will return when discussing our respondents' background): those relatively few women who succeeded in their science career paths to the point of winning a prestigious postdoctoral fellowship may have had an especially bountiful amount of human capital—advantages in their personal background or their training—that helped them overcome (or compensate for) the obstacles and disadvantages typically awaiting women on their path toward becoming scientists.

Focusing on a group of scientists who were judged to be of high ability at the postdoctoral stage should help us decide whether the glass ceiling or the threshold is the more appropriate notion for describing the later careers of talented women scientists. Of course, we acknowledge that our chosen research site also entails certain limitations. Our sample is not representative of the general population of doctoral scientists in the United States. Our findings about gender differences and gender similarities, therefore, may not apply to scientists in general. Furthermore, our population is relatively limited in size and at the same time highly diverse in relevant characteristics, such as academic age and field, so that these characteristics could not be controlled for to a very precise degree.[2] This study is not intended to compete with the more usual studies of representative samples; it aims at complementing them by providing an in-depth look at an interesting subgroup of American scientists.

The Postdoctoral Fellowship As a Stage in Science Careers

Compared with the well-established hierarchy of professorial ranks, the postdoctoral fellowship is a relatively recent addition to the academic career path in the sciences. Although the first postdoctoral fellowships in the United States were offered as early as 1876 by Johns Hopkins University, they became more common only in recent decades. In the 1950s, for instance, both the NSF postdoctoral fellowship and the NRC postdoctoral associateship (which our study will be examining in detail) were instituted. Since the early 1960s, a major increase in the proportion of doctorate recipients in the sciences who planned postdoctoral studies occurred—from about 10 percent in 1960–1961 to about 30 percent in 1978–1979 (NRC 1981, 80). As a result of this increase, the postdoctoral fellowship was, by the early 1980s, considered a standard stage in an academic science career. "The postdoctoral appointment is an accepted feature of the research university," an NRC-sponsored study concluded in 1981 (NRC 1981, 40).

There are, however, important differences by academic field. A postdoctoral fellowship is still much less common in the social sciences than in the natural sciences, although the social sciences also experienced the expansion of postdoctoral fellowships. Whereas 63 percent of the 1978–1979 doctoral graduates in the biosciences and about half of those in physics and chemistry planned postdoctoral study, less than 10 percent of the social scientists had such plans (NRC 1981, 81,112).[3]

What makes a global assessment of the role of a postdoctoral fellowship in a scientific career difficult is that the fellowship can serve several functions (Reskin 1976). The principal difference is between a qualifying and a holding function. The qualifying function of postdoctoral fellowships is to "provide for continued education or experience in research usually, though not necessarily, under the supervision of a senior mentor" (NRC 1981, 11). The postdoctoral fellowship is seen as a transitory stage that prepares for a career as a research scientist. The postdoctoral fellowship offers the young scientist the opportunity to establish a strong publication record and the professional connections that facilitate later career moves.

A postdoctoral fellowship has become a prerequisite for an academic career in first-rate institutions. "For the young physicist or chemist interested in a position in a major research university, 1 or 2 years experience as a postdoctoral may be considered almost essential" (NRC 1981, 102). Former fellows have more publications, on the average, than their cohorts without postdoctoral training (Reskin 1976); and they work more often at major research universities. In 1979, the fifty-nine largest research institutions reported that of the five assistant professors in chemistry whom they had hired most recently, 88 percent had held a postdoctoral fellowship (NRC 1981, 82). Thus, in many fields, there is hardly an alternative for those who aspire to joining the faculty at a major university. They must take a postdoctoral fellowship because they cannot become an assistant professor straight away.

According to the meritocratic principles, which are widely accepted as the normative allocation mechanism in science, postdoctoral fellowships, especially if they have significant career-enhancing functions, should be awarded to those with the best performance in graduate school. In this respect, the postdoctoral fellowship is a honorific reward for previous excellence. A completely different aspect of the fellowship is its holding function. Here it provides a job when other scientific employment is not available. In this different scenario, it is not the best but the marginal young scientists whom the labor market forces into the postdoctoral fellowships when they cannot obtain faculty positions. For these marginal scientists, the postdoctoral fellowship is likely to constitute a "holding pattern" (NRC 1981, 91). Rather than a short transitional phase before moving up to faculty positions, the postdoctoral status may be prolonged for the lack of other employment and may then resemble more permanent nontenure-track positions, such as research associate.

Reskin (1976) emphasized that postdoctoral fellowships vary widely in a number of important dimensions—and, consequently, in in the extent to which they fulfill the functions we have discussed. By concentrating on prestigious national fellowships, we reduce in our study the functional ambivalence surrounding postdoctoral fellowships. Following Reskin (1976), we assume that the recipients of those prestigious fellowships are the most likely among the group of all postdoctoral fellows to have taken the fellowships for their qualifying function rather than for their holding function. We do, of course, investigate to what extent even the recipients of very prestigious postdoctoral fellowships chose them as a holding pattern and whether there were gender differences in this respect.

The Changing Situation of Women in Science

To understand the composition of our female study population, one needs to remember that the representation of women in science underwent decisive, tectonic changes during the past three decades. After certain structural barriers impeding women's careers in science, as well as in other fields, were largely eroded in the early 70s (partly in response to national legislation), women made considerable advances in employment in science. As Hornig (1984, 50) pointed out, "somewhere around 1970 a real discontinuity appeared in all of the educational and employment data involving women." This watershed can also be detected in our population of former postdoctoral fellows. Whereas, for instance, less than 5 percent of the NSF postdoctoral fellowships were awarded to women before 1975, women constituted about 20 percent of NSF awardees in 1975 and later. The career outcomes for former women fellows also improved. Table 1.1 presents our findings on the current mean academic ranks for five-year Ph.D. cohorts of men and women NSF and NRC fellows in our sample who currently work part or full time as scientists in academe.

As these data show, men academic scientists hold higher ranks than women, on the average, in every cohort except for the youngest

Table 1.1. Mean Ranks of Ph.D. Cohorts in Our Sample (without Nonacademic Scientists)

Year of Ph.D.	Men	Women
1955–1959	3.7 (0.6) N=13	2.7 (1.5) N=6
1960–1964	3.8 (0.5) N=34	3.6 (0.9) N=11
1965–1969	3.8 (0.6) N=53	3.3 (0.9) N=12
1970–1974	3.2 (1.1) N=31	3.0 (0.9) N=9
1975–1979	3.0 (0.9) N=107	2.5 (0.8) N=38
1980–1984	2.3 (0.8) N=86	1.9 (0.8) N=29
1985+	1.5 (0.5) N=10	1.7 (0.6) N=3

NOTE: Parentheses contain standard deviations. Academic ranks were assigned numbers as follows: 1 = nonprofessorial positions; 2 = assistant professor; 3 = associate professor; 4 = full professor.

one. The oldest cohort stands out as different from the rest. In general, the average rank increases with time since doctorate; but the women who graduated in the second half of the 1950s hold an average rank that is almost a full step lower than that of the next younger cohort of women (and one full step lower than that of their male counterparts). Because there are only six women in the 1955–1959 cohort, this finding by itself carries little weight. But it certainly seems plausible that the gender difference in this cohort reflects the widespread earlier career pattern according to which a disproportionate number of women scientists were allocated marginal academic positions. In contrast, younger women who obtained their Ph.Ds in 1960 or later appear to be more similar, although not equal, to men as regards average academic rank attained.

Because there are so few women in academe in the pre-1960 Ph.D. group who received NSF or NRC postdoctoral fellowships, the large majority of the women in the Project Access sample belong to that later pattern. Thus, this sample mainly consists of what one might term the "takeoff" generation of women scientists. Most of these women scientists had not even started their professional careers or were still in their early career stages during the watershed years around 1970. Compared with the few isolated pre-take off women scientists, women scientists of this generation were more numerous (although the increase differed by disciplines), and their careers have been more similar to those of their male cohorts. Nevertheless, some gender differences in this generation's careers remain, and exploring them is one of the goals of this project. These differences may not be as blatant as the ones faced by earlier generations of women in science, but they may also be larger than those of the coming generations of women scientists who are only now entering their professional careers. Thus, this takeoff generation of women scientists reflects the profound changes and transitions in the situation of women in science.

Our respondents have successfully mastered the earlier phases of the science pipeline up to the doctorate and accumulated their share of advantages, first of all the reward of a prestigious postdoctoral fellowship. Thus, the women in this sample have already overcome any early obstacles they may have faced in terms of family environment, school, college, or graduate school. Therefore, our sample emphasizes the obstacles at the later career stages (which relatively few women experience because many have already dropped out at an earlier stage). In terms of internal versus structural obstacles, the latter will receive heightened attention. If the traditional gender-role socialization is still in force in society at large, it has certainly not been able to keep these women from pursuing a science career all the way to the end of formal education. But we will also look for reflections of traditional gender-socialization patterns, even among our atypical women respondents.

Samples and Methods

We have already indicated that, in our view, deep changes in the social reality of women in science necessitate a corresponding change in the method of studying career outcomes and obstacles. In the first phase of Project Access, we analyzed a great deal of quantitative data that resulted from a detailed questionnaire about the careers of former postdoctoral fellows. Collection and analysis of such statistical data are without doubt necessary to obtain an overview of broad societal phenomena. By themselves, however, they hardly give more than a simplified picture of the complex and idio-syncratic career paths of our respondents.

We became increasingly aware that a quantitative approach is insufficient if one wants to capture the richness of individual career paths in their subtleties—which seemed to become more and more important. This dissatisfaction with the limitations of purely quantitative methods is shared by other researchers in the area—for instance, Cole (1987, 369): "Typical quantitative methods can estimate rather well features of the formal aspects of citizenship in science. . . . However, these techniques, at this stage of their development, do not allow us to measure adequately other, informal aspects of citizenship." Consequently, Project Access embarked on a more qualitative second phase, which was based on open-ended, face-to-face interviews with a subgroup of our respondents.

This monograph, then, attempts to understand a complex reality by drawing on different methods, triangulating the situation of women in science from quantitative and qualitative vantage points—ranging from a multiple-choice questionnaire, to codings of open-ended questions, to the former postdoctoral fellows' own words. By doing a qualitative study after a quantitative one, Project Access reversed the conventional sequence of social-science research, which often proceeds from exploratory studies of a more qualitative kind to quantitative studies. Our sequence allowed us to investigate the fine structure of phenomena identified by quantitative analyses. Needless to say, we hope that our qualitative phase is not an ending but stimulates further research of the quantitative or qualitative kind.

The presented research is based on two sets of data: a larger set of questionnaire data from 699 former postdoctoral fellows, and a smaller set of two hundred open-ended interviews.

Questionnaire

Our set of 699 former postdoctoral fellows comprises men and women who received a National Science Foundation (NSF) fellowship from the inception of the program in 1952 through 1985 or a National Research Council (NRC) associateship from its start in 1959 through

33

1986. The respondents include 460 former NSF Postdoctoral Fellows (99 women and 361 men) and 239 former NRC Postdoctoral Associates (92 women and 147 men). This was the most extensive sample of former NSF and NRC postdoctoral fellows that could reasonably be obtained. The NRC, which keeps an address database of its former associates, made available the addresses of all former associates who indicated their willingness to participate. The NSF gave us the names of all their former fellows, whom we ourselves then had to locate, chiefly by contacting the alumni offices of their colleges or graduate schools.[4]

Our data were collected between 1987 and 1990 from a lengthy questionnaire that was sent out to those former fellows who could be located. Repeated attempts were made if the first mailing did not result in a response. Of the former postdoctoral fellows who were contacted, the response rate was 60.6 percent for former NSF fellows (62.1 percent for men and 55.6 percent for women), and 82.1 percent for former NRC associates (81.7 percent for men and 82.9 percent for women). Our respondents' mean year of receiving the doctorate was 1975 (men: 1975; women: 1976). The mean birth year of the respondents was 1946 (men: 1946; women: 1946).

Among our respondents, slightly fewer men than women were U.S. natives (87.6 percent versus 90.5 percent), and fewer men than women were naturalized American citizens (3.6 percent versus 6.3 percent). Information about racial and ethnic heritage was available only for our NSF respondents. In this group, all the women and 97.5 percent of the men were white, eight men (2.2 percent) were Asians or Pacific Islanders, and one man was black (0.3 percent). This racial composition differs markedly from the racial makeup of the current population of science graduate students, substantial numbers of whom are nonwhite. On the one hand, of course, our respondents reflect the predominantly white graduate-school cohorts of past decades. On the other hand, NSF requirements of U.S. citizenship (or permanent residence) for its fellows may have eliminated some potential nonwhite applicants. Among the white former NSF fellows, two women and fifteen men were of Hispanic heritage. Thus, Hispanics form the largest minority (4.7 percent) in our sample; but fourteen (82.4 percent) of the seventeen Hispanic fellows traced their ancestry to Spain, not to countries of Central or South America.

34

Reflecting a general pattern among scientists, a higher proportion of women than men respondents works in the biological (46.1 percent versus 29.7 percent) or social sciences (17.8 percent versus 8.1 percent), whereas a lower proportion of women works in the physical sciences, mathematics, and engineering (36.1 percent versus 62.2 percent).

To form a general impression of the representativeness of our sample, we compared the distributions of three key variables (gender, year of fellowship, and academic field) in the active population, in the contacted sample, and among our respondents. Information on these variables was available from NSF records, regardless of whether a person responded or not. The active population are all those former fellows who fall within the scope of our study. The contacted sample are those who were considered to have been contacted by mail (that is, all those who were sent a questionnaire minus those relatively few whose mailings were returned by the Post Office as undeliverable). Inspection of the mentioned key variables did not indicate any major bias among our respondents as compared to the active population or the contacted sample (see Appendix).

Face-to-Face Interviews

After a period of developing, testing, and revising an open-ended interviewing procedure, one interviewer traveled extensively from summer 1989 through fall 1990 in order to conduct two hundred face-to-face interviews with former postdoctoral fellows all across the United States. The interviews typically lasted from two to three hours each and were usually conducted in the interviewees' home institutions or homes. The interviewees were first asked to narrate their whole career path in their own words and were then asked the open-ended questions one by one.

A subset of twenty-four interview transcripts (twelve males and twelve females) were thoroughly studied to distill themes of typical responses and develop a coding scheme. Thus, the categories in the coding scheme emerged from the interviews themselves rather than being superimposed on them from some prior information or theory. Then all interview transcripts were coded according to the coding scheme. Five coders participated in the coding; the overall Cohen's kappa was 0.698.

The interviewees include 108 women and 92 men who had been National Science Foundation (NSF) postdoctoral fellows (114), National Research Council (NRC) postdoctoral associates (51), Bunting postdoctoral fellows in the sciences or engineering (28), or Bunting finalists in these fields (7). Different considerations entered the selection of interviewees from our larger sample of former recipients of these prestigious postdoctoral fellowships. On the one hand, we attempted to contact as many former fellows as our limited budget made possible. On the other hand, we wanted to match men and women in the sample by type of current position, academic age, and academic fields. The resulting sample is a tradeoff: a larger sample that is not perfectly matched, but is matched to a substantial degree.

Of the respondents, 58.0 percent (men: 57.6 percent; women: 58.3 percent) currently work in academic science; 30.5 percent (men: 31.5 percent; women: 29.6 percent) work as scientists outside of academe; and 11.5 percent (men: 10.9 percent; women: 12.0 percent) have left research science. The mean year of receiving the doctorate was 1974 for the men and 1971 for the women. In terms of fields, more than a third (37.5 percent) of the respondents are in the biological sciences (men: 38.0 percent; women: 37.0 percent). Men are overrepresented in the physical sciences, mathematics, and engineering (men: 50.0 percent; women: 32.4 percent) and, conversely, underrepresented in the social sciences (men: 12.0 percent; women: 30.6 percent). As to geographical location, a large number of interviews (sixty-eight) were conducted in New England, reflecting both a geographical concentration of former postdoctoral fellows in northeastern states and easy accessibility from the Massachusetts-based Project Access. Thirteen interviews were conducted in the state of New York, and forty-seven took place in the Middle Atlantic region. Another geographical focus was the West Coast region, with forty-one interviewees. The remaining interviewees lived in the Midwest (eighteen), the South (seven) and the Mountain States (six).

This monograph not only presents results from the questionnaire and from the coding of the open-ended interviews, but adds a great number of passages from the interviews. One of the key benefits of conducting open-ended interviews is that respondents talk in their own words. Thus, quotations give more detail and immediacy than summary statistics are able to do.

Both questionnaire and interview data have been archived and can be accessed for scholarly research at the Henry A. Murray Research Center of Radcliffe College.

Plan of the Book

The first issue we addressed was, Do the gender disparities that have been often observed in the general population of American scientists also hold among our specially selected sample? In other words, is there a threshold or a glass-ceiling effect among our group of particularly promising scientists? We first turned to basic outcome measures, such as leaving research science and academic rank.

After finding a glass ceiling, at least in some disciplines, we turned to explaining it. Within a Mertonian framework, it is the disparity between the genders—not gender equality—that needs explanation. Instead of immediately attributing career disparities to discrimination, we followed a process approach. The first step was to control for performance criteria. Then we moved into the area of human-capital characteristics (personal background, graduate school, and postdoctoral variables) to check to what extent the process of the accumulation of advantages and disadvantages was influenced by gender. Next we used the dichotomy of deficit model versus difference model as the theoretical framework for a more detailed description of scientists' career paths and obstacles, drawing heavily on the scientists' own perceptions. The focus was on both lingering structural obstacles and gender differences in the style of doing science. Finally, we discussed the science-external issue of marriage and parenthood.

2

OUTCOMES: GENDER DIFFERENCES IN SCIENTIFIC SUCCESS

WE NOW TURN TO examine gender differences in basic outcomes for our 699 former NSF and NRC postdoctoral fellows, including present employment (still in science or outside of science), field of scientific employment (in or outside of academia), and the job characteristics of those who work within academic institutions. Excluded from the following discussion of career outcomes are three retirees—two men and one woman.

Within our theoretical framework, we examine if the female former fellows' careers proceeded according to the threshold or the glass-ceiling pattern. We look at the basic outcome parameters to get a first indication of whether fundamental gender disparities exist in our sample and what their dimensions are. In a pragmatic approach, this look at basic career outcomes lets us gauge the severity of the problem of women's science careers in our special sample. If we find gender disparities, even in our select group of promising scientists, we will examine the gender differences in greater detail, also using qualitative data. If we discover that our women respondents' careers follow, on the whole, a threshold pattern, we will not be able to declare that the problem of women's science careers has gone away. But such a result would help locate the problem more precisely. It would direct future research toward other subpopulations of American scientists where gender disparities might be concentrated.

Present Employment

Although economic necessity and changes in social values have weakened the strict gender division of labor along workplace-household lines, wives' careers are still often considered only a source of supplementary family income. By the same token, part-time employment is more acceptable and widespread for women than for men, particularly because of women's traditional family obligations. Even highly educated and skilled women, such as scientists, are underutilized in the scientific labor force (Hornig 1984, 51; Vetter 1984, 61,71).

In the general population of American scientists, it has consistently been found that a higher proportion of women than men is involuntarily unemployed (NSF 1990, 8; Vetter 1987, 7). This was also found in our sample; but the difference was not statistically significant, owing to the small numbers involved (two unemployed women and three unemployed men) (see table 2.1).

Our sample of former postdoctoral fellows also resembles the general population of American scientists as far as part-time and full-time employment is concerned (NSF 1990). Among our respondents, a lower proportion of women than of men was full-time employed, and the likelihood of working part time was significantly higher for women than for men.

Sector of Present Occupation

Of all the former NRC and NSF fellows, only a small minority had left research science, with men and women in similar proportions (men: 8.5 percent; women: 10.0 percent; $p = 0.5361$). A logistic regression controlling for fellowship, field, and year of doctorate also did not reveal any significant gender difference in this respect. Field differences, however, were evident. Social scientists were most likely to leave research science, whereas respondents in the physical sciences, mathematics, and engineering (PSME) were most likely to stay.

Table 2.1. Employment Status (in Percent)

	Women	*Men*
Postdoctoral or student	12.2	7.8
Full-time employed	78.8	87.8‡
Part-time employed	6.3	1.8†
Unemployed (seeking employment)	1.1	0.6
Other	1.6	2.0
N	189	502

NOTES: Retirees excluded. Table shows two-tailed t-tests of gender difference here and in subsequent tables, unless noted.
†$p < 0.05$
‡$p < 0.01$

Most former fellows who left science work in nonscientific capacities in business and industry, government, or educational institutions. Within these occupations, no clear gender trend could be discerned, which may be due to the small size of this group (twenty women and forty-three men). The clearest indication of a gender difference appeared, as one might expect, in the category "homemaker/child care." None of the men was a homemaker, whereas three female former fellows (15 percent of female nonscientists) gave homemaker as their primary occupation.

For those respondents who are still active research scientists, we examined whether they worked in academe or in nonacademic science. An often-observed gender division of labor among persons with doctorates encompasses different choices of employment sectors. Researchers have found that women scientists are more likely than men to work in the academic sector (Centra 1974; NAS 1979, 57). To some extent, this sectoral difference may be due to gender differences in the choice of academic discipline. After controlling for field, the sectoral gender difference is attenuated in the general scientist population (NRC 1983, chap. 4, 1). In our special group of former postdoctoral fellows, men were *more* likely to work in academe than women (69.8 percent versus 58.5 percent; $p = 0.0074$).

We performed a logistic regression of a dichotomous variable that distinguishes between scientific employment within and outside the academic sector on fellowship, fields, years since doctorate, and gender. The strongest predictor in this regression was fellowship. Most former NSF fellows embarked on careers in the academic sector, whereas the majority of former NRC Associates worked outside academia—that is, mostly for the federal government or in big non-university research laboratories. The proportion of former fellows in our sample who worked in business and industry hovered around 10 percent to 15 percent regardless of gender or fellowship. Another significant predictor of employment sector is years since doctorate; older scientists were more likely to work within academia. The gender coefficient indicated a higher likelihood for men than for women to work in academe, but the gender difference became nonsignificant in the multivariate approach ($p = 0.1960$). The finding that our men respondents were well represented in academic science (even though they were somewhat predisposed to nonacademic science in the general population) is consistent with Reskin's (1976) finding that a postdoctoral fellowship increased the likelihood for men chemists, but not for women, to work in the academic sector.

Academic Rank

Before we discuss academic rank and, later, departmental prestige as key indicators of how women fare within the social system of science, we should note that, strictly speaking, academic rank does not belong to the social system of science but to the social system of universities (or of academe). First, the social system of universities also covers nonscience fields, such as the humanities and various professions (medicine, law, business, and so on). Second, in addition to scientific research, universities have other objectives, notably teaching. Strictly speaking, universities are committed to academic values rather than scientific values, and academic rank is a reward for academic merit rather than scientific merit.

Nonetheless, academic rank is a crucial resource for those in the social system of science. Salary, office space, and various research facilities that come as benefits of academic rank provide material preconditions for scientific activity. In addition, few academic institutions would disregard scientific merit completely in determining academic merit; and the more renowned the university, the more it values contributions to scientific knowledge—in other words, the larger the overlap between academic and scientific criteria of merit. Because the social system of academe and the social system of science are closely linked, academic rank promotion is often viewed as "a form of recognition bestowed by the scientific community" (Long, Allison, and McGinnis 1993, 703)—that is, as a reward within the social system of science.

As considered in this book, the four main steps of the academic hierarchy are nonprofessorial positions, assistant, associate, and full professorship.[1] The nonprofessorial positions include jobs such as research associate, research scientist, lecturer, and instructor. They often confer marginal status to scientists, especially when they are permanent positions (Rosenfeld 1984, 95). The four-step hierarchy of rank is, of course, an ordinal not an interval scale. Nonetheless, researchers have commonly treated it as a quasi-interval variable to facilitate its use in statistical regressions (see Ferber and Loeb 1973; Rosenfeld 1984), and we follow this usage in our outcome variable of rank. We will complement the examination of this overall rank variable with logistic regressions of individual rank stages.

Reflecting the fact that in most fields of science men have numerically predominated in the past, women's underrepresentation is predictably most severe in the senior positions of the academic hierarchy. Nationwide, the largest group of women scientists in academe were assistant professors in 1981, whereas the largest group of men were full professors (NRC 1983, fig. 4.2). In a situation where most of the women entered science relatively recently, it is necessary to control for academic age when making gender comparisons. In our sample, academic age is fairly well balanced; the men's average academic age is only about one year higher than the women's. Still, the largest group of the men were full professors (41.5 percent), whereas the largest group of the women were associate professors (29.6 percent); and only 23.2 percent of the women had achieved the position of full professor.

Regression analysis showed, as one might expect, that the time period since the award of the doctorate predicted rank most strongly, explaining 36.96 percent of the variance in rank by itself. Working in

Table 2.2. Regression Models of Rank on Years since Doctorate, Fellowship, Fields, and Gender

	1	2	3
Years since doctorate	0.08§	0.08§	0.08§
Fellowship	0.34§	0.28‡	0.99§
Biology	−0.39‡	−0.41‡	−0.25‡
PSME	−0.05	−0.13	0.39
Gender		−0.27‡	0.01
Gender × years since doctorate			−0.00
Gender × fellowship			−0.53‡
Gender × biology			0.61†
Gender × PSME			−0.48*
R^2	0.4125	0.4241	0.4689
N	422	422	422

NOTES: Unstandardized regression coefficients. Rank: 1 = nonprofessorial positions; 2 = assistant professor; 3 = associate professor; 4 = full professor. Fellowship: 0 = NRC; 1 = NSF. Gender: 1 = male; 2 = female.
*$p < 0.10$
†$p < 0.05$
‡$p < 0.01$
§$p < 0.001$

biology was associated with a lower rank compared with working in other science fields. Controlling for years since doctorate, fields, and fellowship, we noted that gender had a significant effect (table 2.2, regression 2). All these control variables being equal, a woman's predicted rank was more than a quarter lower than a man's. When interactions between gender and the other predictors were added to the model, the gender main effect disappeared, but the interaction of gender with biology became significant (table 2.2, regression 3). This interaction indicates that the women's disadvantage in academic rank was concentrated in the nonbiological science fields, notably in the physical sciences, mathematics, and engineering, where the women's average rank was much lower than that of comparable men. This finding points to the importance of academic fields: the gender gap appears to be smaller in biology than in the other sciences. This will become a recurrent theme in our study.

Among our biologist respondents, women's progress through the ranks was roughly similar to men's. Greater gender equality in biology was, however, paralleled by lower overall ranks of biologists than of other scientists. This leads to the question of whether the two findings are connected. One might speculate that academic career schedules in biology are more standardized than in the other sciences. In other words, advancement policies in biology might be less accommodating toward high fliers or "geniuses" who might be promoted to top academic positions at a very early age in other disciplines (mathematics, in particular). If women scientists, for familial or other reasons, were less likely to pursue optimal career trajectories, a standardized advancement schedule would give them more opportunities to catch up and compensate for temporary setbacks or obstacles. There was also a gender interaction with the type of fellowship; men appeared to benefit more than women from an NSF fellowship. The NSF postdoctoral fellowship launched males in particular onto the fast track toward academic success, whereas for female recipients as a group, the advantage of such a fellowship may have been partly offset by various obstacles.

Tenure

We now turn from the broad consideration of academic ranks to two critical junctures: receiving tenure and getting a tenure-track position. Typically, tenure status and academic rank are correlated. Assistant professors are mostly on tenure track, and associate and full professors are usually tenured; but this correlation is far from perfect (NAS 1979, 62). Some of the most prestigious universities, for instance, tenure only full professors.

In the relation between academic rank and tenure status, a gender anomaly has been observed. Nationwide, the proportion of women assistant professors outside the tenure track has been found to be almost twice as high as the corresponding figure for men assistant professors (Chamberlain 1988; NRC 1983, chap. 4, 15). Our data, by contrast, did not show such a pattern. Eighty-two percent of the women assistant professors were on tenure track, compared with 84 percent of the men assistant professors. We did, however, find a

45

larger gender difference in the nonprofessorial category of instructor, lecturer, or other. Here, men respondents were more than twice as likely as women to have tenure or be on tenure track (16.7 percent versus 5.9 percent; $p = 0.26$), although the difference did not reach significance. Nonetheless, there was at least a hint that in these marginal positions, men obtained a higher level of job security than women.

Among the general population of doctoral-level scientists and engineers who were employed in universities and four-year colleges in 1987, women were much less likely to be tenured (36 percent) than men (60 percent) (NSF 1990, 7). In the biological sciences, 41 percent of the women, compared with only 15 percent of the men, were neither tenured nor on tenure track in 1985 (Vetter 1987, 7). Among those scientists in our sample who now work primarily in the academic sector, 67.1 percent of the men but only 48.0 percent of the women had received tenure ($p = 0.0005$).

In a logistic regression, controlling for years since doctorate, fields, and fellowship, we noted that being female appeared to decrease the chance of having tenure, although this was not quite statistically significant ($p = 0.1172$). Again, the picture differed substantially by field. A strong interaction between biology and gender indicated that the tenure situation was favorable for women in biology; women achieved tenure at equal or even at higher rates compared with men. For instance, in the group of younger biologists (who received their doctorate in 1978 or later), the proportion of women with tenure was twice as high as the corresponding proportion of tenured men. By contrast, women were at a substantial disadvantage, compared with men, in the other fields.

Tenure plus Tenure Track

We now combine tenure and tenure-track positions to address the question of whether a higher proportion of women than men was permanently marginalized in nonprofessorial positions or whether the women's careers, on average, were similar, though sloweddown, versions of the men's careers. In the latter case, women's underrepresentation in tenured positions would be compensated by their overrepresentation in tenure-track positions so that gender differences would diminish on the combined measure of tenure plus tenure-track positions.

Controlling for year of doctorate, fields, and fellowship, we noted that the negative gender main effect was too small to reach significance ($p = 0.2107$). There were, nevertheless, interaction effects of gender with fellowship and gender with fields. Again, men were better able to capitalize on NSF fellowships, perhaps because they forged better connections during their fellowships. Women in PSME had more difficulties in obtaining tenure-track or tenured positions than women in the other fields. Thus, in PSME we observed the permanent marginalization of women, whereas in the other fields a lag in women's rank achievement, if there was any, indicated a slowed-down career trajectory that nonetheless progressed toward the same final destination (full professorship).

Departmental Prestige

For American academics, it not only matters what rank one holds but also where one works. The rank hierarchy exists uniformly across all academic institutions; but in a second dimension of academic job quality, scientists are stratified by the prestige of the department or the institution to which they belong. A nationwide assessment of research-doctorate programs (Jones, Lindzey, and Coggeshall 1982) provided a measure of the prestige of the departments with which our academic scientists were affiliated. The ratings from this study are based on a large reputational survey in which American faculty members evaluated, among other things, the quality of faculty in research-doctorate programs in their disciplines. The ratings were standardized across disciplines, with a mean score of 50 and a standard deviation of 10.

Not all academic institutions of higher learning received ratings, mainly because the assessment focused on research-doctorate programs unavailable at colleges catering mostly or exclusively to undergraduates. The great majority of the academic scientists in our sample worked at institutions with research-doctorate programs, which is not surprising given our respondents' high scientific promise at the postdoctoral stage. But 25.3 percent of the female scientists and only 15.7 percent of the male scientists in our sample were located at institutions without research-doctorate programs (p = 0.0475). The difference was marginally significant in a multiple regression including years since doctorate, fellowship, and fields. This remained the only gender difference in affiliational prestige we found. Among the respondents at rated institutions, the mean departmental prestige rating was above the national average equally for both genders (men: 56.1; women: 56.7). And the proportions of men and women respondents working in the top 15 percent of departments were also similar (men: 28.0 percent; women: 30.1 percent).

Thus, a disproportionate number of women respondents worked at academic institutions without strong graduate programs; but the group of women respondents who did work at research universities was not at a disadvantage, compared with men, in terms of affiliational prestige. They were equally represented in the top departments. But did they also hold equal ranks at these institutions?

Rank and Affiliation

Although rank and prestige of department have been identified as two different aspects of academic job quality, "too few studies combine data on institutional affiliation with data on organizational rank" (Zuckerman 1987, 131). We wondered, first, whether academic position and departmental prestige are related (most likely in the form of a tradeoff) and, second, whether a tradeoff relationship between academic position and affiliation exists for women scientists in particular, as some studies indicate (Rosenfeld 1984). To put it more technically, we added departmental prestige—a dummy variable distinguishing whether or not a respondent was affiliated with one of the departments in the top 15 percent—as an independent variable to regressions of academic position. In these regressions we were interested in both the main effect of departmental prestige and its interaction with gender.

Departmental prestige appeared unrelated to academic rank in main-effect models, indicating that these are indeed two relatively independent dimensions of academic job quality. After we included gender interactions into the model, however, a more complicated pattern emerged. As a group, women respondents experienced a considerable rank disadvantage at the most prestigious departments. In other words, women "paid" to a significant degree for working at more prestigious departments with a loss in rank, whereas men did not experience such a "tradeoff." It should be emphasized that we are dealing here with *aggregate-level* data. Thus, the "trade-off" applies to women as a group. It does not suggest that individual women scientists commonly faced an actual career choice of high rank in a less prestigious department versus low rank in a more prestigious department.

We shall now move from considering the whole ladder of academic ranks as one variable to looking more specifically at important rank and status differences. Multiple measures of academic position reveal the fine structure of the relationship between academic position and affiliation. Main-effect logistic regression models that included our dummy of affiliational prestige indicated that significantly fewer respondents held tenured or tenure-track positions at high-prestige departments than at less prestigious departments. Only in this aspect did we find a tradeoff between rank (or status) and departmental prestige that applied equally to women and men. This tradeoff may be easily explained as a result of the structure of American academic science. Research requires a considerable number of postdoctoral fellows, research associates, and other doctoral-level researchers. Research activities are concentrated at high-prestige universities. Thus, high-prestige departments tend to have a comparatively large proportion of nonprofessorial positions for scientists.

The interaction between gender and departmental prestige plays a role for predicting rank, as we noted, but not for predicting tenure or tenure plus tenure track. In other words, women at highly prestigious institutions were not so much handicapped, relative to men, in terms of tenure or tenure-track attainment as they were in terms of academic rank. These results suggested that the gender-prestige interaction was particularly strong at the level of full professor. An examination of the gender differentials at the rank of full professor supported this suggestion. As regards the attainment of full professorship, the gender disparity was larger at top departments than at others ($p = 0.0295$). Again, biology appeared to be a favorable place for women scientists. Three-way interactions between gender, departmental prestige, and fields were not significant; but in regressions for biologists alone, the tradeoff effect was nonsignificant.

Future research should focus on the top rank at the top universities, where the women's tradeoff between rank and affiliation was the strongest, in order to determine if this gender barrier is only temporary or permanent. A clue to the persistence of the glass ceiling at the top universities may lie in the tenure variable. Once men and women academics have achieved tenure status, they can expect to reach the top academic position of full professor unless something goes drastically wrong. Women who have reached tenure (typically coterminous with promotion to associate professor) have thus all but secured the top outcome in terms of rank.[2] At worst, their careers might be slowed-down versions of men's careers, but eventually they will catch up. In a regression of tenure status, there was an interaction between gender and affiliation in the predicted direction, but it was nonsignificant ($p = 0.3328$). Women may, if anything, be slightly disadvantaged in terms of tenure attainment at the most prestigious departments. This result suggests that the observed gender disparities at the level of full professor, controlling for academic age, will attenuate over time as the tenured women scientists with slightly slowed-down career paths catch up; but it might not entirely disappear for a while.

The main objective of this section was to determine whether the concept of the glass ceiling or of the threshold would better describe the careers of women who had received prestigious national postdoctoral fellowships—in other words, whether the career paths of the former fellows would diverge in later stages in a gender-specific way. Even among our very promising scientists, women appeared to be underutilized in the scientific labor force compared with men. Women however, as a group, showed remarkable tenacity in terms of remaining research scientists. More women than men may work part time, but only a small minority of women respondents, similar in size to the men's proportion, had left research science altogether. Furthermore, in contrast with findings for the general population of scientists, women in our sample were not overrepresented in the academic sector of science.

As regards gender stratification within academic science, the importance of academic field was striking. The biological sciences proved much more hospitable to women in terms of career advancement than the other sciences did. Although main-effect models regularly found women scientists at a disadvantage compared with men, models with gender interactions could locate this disadvantage more precisely in the nonbiological sciences. Among the biologists, the women's careers proceeded in roughly the same pattern as the men's. The women academic biologists in our sample appeared to have overcome a threshold. By contrast, the glass-ceiling notion seems more appropriate in the other sciences. In spite of any affirmative action efforts that academic institutions may have undertaken in the physical sciences, mathematics, and engineering, the relatively few women respondents in these sciences have, as a group, had less successful academic careers than the men. In other words, women encountered the fewest career obstacles in a field where they are represented in relatively high numbers. This finding is consistent with Kanter's (1977a, b) view that rare representatives of a social group, called tokens, face particular difficulties. Although this appears to be the case in the physical sciences, mathematics, and engineering, women biologists may have reached a critical mass that inactivates gender stratification and segregation. One may speculate that the relatively long and strong tradition of women in biology, as compared with other natural sciences, has contributed to the reduction of the gender gap in that field. Women in biology may not be strangers as they are in other disciplines.

Women were slightly underrepresented at institutions with research-doctorate programs. Nonetheless, among the scientists who worked at such institutions, women were affiliated with departments of similar prestige, compared with men. But for this group of women, higher departmental prestige was connected with lower rank, whereas men did not experience such a tradeoff relationship. This glass-ceiling effect appeared strongest at the level of full professorship at highly prestigious departments.

Career patterns also differed by the kind of postdoctoral fellowship. The NRC associateship program is geared toward a career in nonacademic science, whereas NSF fellows tend to choose primarily academic science paths. Among those former postdoctoral fellows from both groups who had become academic scientists, the former NSF fellows were more successful than the former NRC associates.[3] Their greater success in academia can be specifically attributed to male NSF fellows; female fellows did not seem to reap similar benefits.

Our results indicate that the patterns of gender-specific segregation and stratification in science have become increasingly complex. There are serious gender disparities in the sciences outside of biology, both in terms of rank and the top positions at top universities. This finding warrants further investigation into the mechanisms that have brought about these gender disparities in science careers— focusing on small differences and subtle mechanisms. But first we have to check whether these disparities can partly or wholly be attributed to differences in scholarly merit.

Publication Productivity

So far, the discussion of the career outcomes of our academic scientists has focused on basic reward dimensions—that is, academic rank and affiliational prestige. The rationale for not including a performance variable as a matter of course was that constructivist scholars (see chapter 1) would reject in principle any concept of objective merit in science. Those who hold an outcome notion of discrimination would be predominantly interested in the outcomes and outcome differentials that we presented.

By contrast, for those with a more Mertonian approach, the concept of performance—problems of its operationalization notwithstanding—is a central aspect of science careers. The basic norm of universalism asserts that rewards should be allocated according to merit or performance. Thus, gender differences in the reward dimension might be justified by differences in the performance dimension. Or, in terms of the process model of discrimination, not every gender disparity in outcomes is by itself proof of gender discrimination.

Therefore, we now turn to the performance dimension. In a Mertonian framework, the mission of scientists is to add to scientific knowledge. Scientists' intellectual contributions to the advancement of their field are the major element in evaluating them as scientists. New scientific knowledge is disseminated in the scientific community mainly through publication. Thus, publication productivity is widely considered the prime indicator of a scientist's performance.

There are obvious problems with a performance measure based on publication counts. The major concern is that publication quantity does not necessarily translate into scientific quality. A mere publication count would, for instance, undervalue a scientist who scrupulously produces a small number of highly significant pieces vis-à-vis a scientist who publishes a large amount of mediocre or marginal material. There have been attempts to modify raw publication counts by factoring in types of publication or journal prestige. Another approach (with problems of its own) is basing scientists' performance evaluations on the number of citations their publications receive.

Among the variety of performance measures, the raw publication count appears to be an attractive indicator of scientific performance for four reasons. It is simple, it correlates strongly with more complex measures,[4] it is easily available, and it is widely used and accepted in the scientific community. In a small study of peer evaluation in biology, we found raw publication productivity to be the strongest predictor of the raters' quality judgments (Sonnert 1991). In this section, we will focus on publication productivity as a central indicator of scientific merit. In particular, we will address the following questions:

Does a gender productivity gap exist in our sample?
Are there differences between fields?
Are there gender differences in the types of publications produced (for example, articles, book chapters, or books)?

Later we will address the issue if publication productivity can really be considered an independent indicator of performance that is then rewarded with certain jobs. In other words, does a good publication record lead to a good job, or does a good job lead to a good publication record? We will also question the appropriateness of the publication-productivity indicator, especially in light of possible gender differences in publication behavior.

Whereas publication productivity is the widely recognized "coin of the realm" in academic science, the importance of publication productivity varies among nonacademic scientists, some of whom may work on highly classified research projects. We therefore limited our focus to academic scientists when examining publication productivity. The publication-productivity data were collected by asking the respondents to send in their bibliographies along with the filled-out questionnaire.

Gender Differences in Scientific Productivity

Studies have repeatedly found that women scientists, on average, are less productive in terms of publication volume than men scientists (Cole and Zuckerman 1984). This finding was replicated in our sample: academic women scientists had an average of 2.3 publications per year (median: 1.8), whereas men had 2.8 publications (median: 2.2). The gender difference was statistically significant by itself ($p = 0.0133$) and after controlling for fields ($p = 0.0379$). The gender gap appeared smaller in the biological sciences (0.18 publications) than in the other sciences (0.89), although this difference between biology and the other sciences did not quite reach significance level (interaction term: $p = 0.1267$).

Because distributions of productivity rates are rarely normal (bell-shaped), averages do not tell the whole story; and a closer description of the frequency distribution of productivity rates is warranted. It has often been observed that many scientists cease to publish in their later career stages and that relatively few scientists write most of the scientific literature (Allison and Stewart 1974; Cole and Zuckerman 1984; Lotka 1926; Price 1963). Reflecting this situation, the typical histogram of the frequency distribution of publication rates for the general population of scientists resembles a decay curve. In fact, in our sample, the productivity distribution has this characteristic shape (figure 2.1). But two comments are in order. First, be-

Percentage

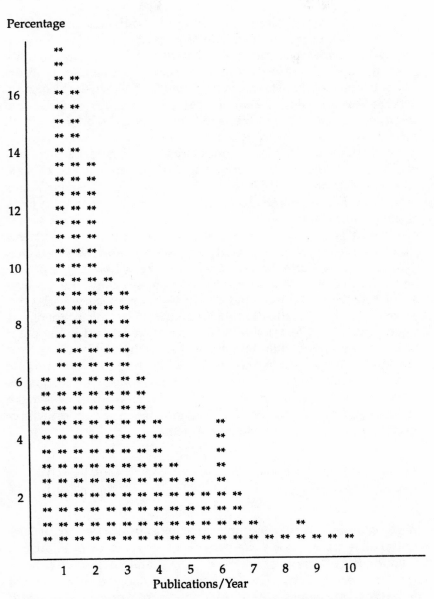

Figure 2.1 Frequency Distribution of Annual Publication Productivity

cause our sample of former recipients of prestigious postdoctoral fellowships contains an unusually large proportion of highly productive academic scientists, the distribution of annual productivity rates, though still skewed, is probably more normal than that of a cross-section of scientists would be. Second, a spike in the histogram occurred around six publications per year, which may indicate a small concentration of prolific scientists. This spike occurred in all three fields.

Researchers found highly productive scientists to be predominantly male (Cole and Zuckerman 1984). This small group of male scientists with huge quantities of publications may be responsible for part of the often-observed gender gap between the productivity rates of male and female scientists. Indeed, in our sample, women comprised only 13.9 percent of the highly productive group of thirty-six scientists with an average annual publication productivity of higher than 5.5 (that is, those who were on the spike or to the right of it in figure 2.1), whereas the proportion of women among the 306 academic scientists who were less productive was twice as high (28.1 percent) (likelihood ratio chi-square; $p = 0.053$). In biology, there was a good number of females with high productivity, but in the other fields hardly any woman was a member of the highly productive segment of the scientists. When the small group of very prolific scientists was disregarded, male and female academic biologists were found to be equally productive. In the other fields, however, the gender gap in productivity decreased but still persisted, even after the very prolific scientists were excluded.

Type of Publication

In our sample, the bulk of publications was in the form of articles (rather than books, book chapters, or other types of publication), although the proportion of articles among total publications varied from 83.7 percent in PSME to 80.2 percent in biology and 66.3 percent in the social sciences. If we consider only journal articles, the gender gap in productivity is again found in all fields. In biology, women on average produced 0.47 articles per year less than the men. In PSME, the difference was 0.50 articles, and in the social sciences, 0.83 articles.

In biology, the gender gap as regards articles per annum (0.47) is more than twice as wide as it was in respect to total publications (0.18). It follows that the women biologists in our sample must have outpublished men in some other types of publication. The women biologists published 0.69 nonarticle publications per year, whereas the men published only 0.41. This difference is statistically significant ($p = 0.0069$). Although the production of books was negligible for biologists of both genders (0.017 books per year for men and 0.016 books per year for women), women biologists appeared to have published marginally more book chapters than their male counterparts (0.41 versus 0.30 book chapters per year; $p = 0.1157$). The same pattern was found for the small category of articles in conference proceedings and similar publications. Women biologists produced 0.18 such publications per year on the average, whereas the men produced only 0.08 ($p = 0.0254$). Finally, in the minute residual category of other publications (including pamphlets, internal papers, letters to editors, and electronic media), women biologists also published more than men (0.08 versus 0.02 publications per year; $p = 0.0387$).

Thus, in biology, women were partly compensating for a lower productivity in terms of articles with a higher productivity in terms of other, perhaps softer publications (that is, less likely to be peer-reviewed), such as book chapters, articles in conference proceedings, and other publications. By contrast, the other fields showed no such pattern. It is not clear whether the situation in biology is due to push or pull factors. On the push side, women biologists may experience more difficulty than men in getting their research papers accepted through the process of journal peer review. Therefore, they might attempt to use other, perhaps somewhat less prestigious, publication channels. On the pull side, the women biologists' tendency toward publishing book chapters and articles in conference reports may indicate that the women are particularly well integrated in collegial networks that facilitate invitations to write book chapters or to attend conferences. Furthermore—and this is certainly an important point that we are going to pursue—women biologists may be more interested than men in publishing substantial synthetic pieces that are too long for regular journals. As regards the residual category of other publications, it is not clear whether the women produced more of these publications or whether they were more diligent in noting these nonstandard items in their bibliographies.

In sum, the often-observed publication-productivity gap between the genders was found even among our sample of former recipients of prestigious postdoctoral fellowships. Among the respondents who now work in academe, the average female productivity was about 80 percent of the average male productivity, or, put differently, the average woman had 0.5 publications fewer per year than the average man did. The width of the gap appeared to vary somewhat by field. The gender gap in productivity was the narrowest in biology because women biologists outpublished men biologists in nonarticle publications (book chapters, conference proceedings, and so on).

Rank, Affiliation, and Productivity

We now look at the reward-type outcome variables of academic rank and affiliational prestige and the performance-type outcome variable of publication productivity in combination. Our data do not lend themselves to deciding the direction of causality between the three aspects of career outcomes, but they let us examine whether they are correlated (table 2.3). We should note that the direction of causality between the three outcome variables has been a matter of considerable debate, partly because the meritocratic nature of the scientific system is at stake. If academic science was a simple meritocratic allocation system, rapid rank promotion and affiliation with prestigious institutions should be the rewards of high publication productivity. If institutional location and rank determined productivity, however, the simple meritocratic view of the social system of academic science would have to give way to a more complex view of science careers as dynamic, interdependent processes in which merit, accumulation of advantages and disadvantages, and self-fulfilling prophecies may all play a role. Indeed, whereas research productiv-

Table 2.3. Partial Correlations among Outcome Residuals

		Affiliation	*Productivity*
All (N=288)	Rank	−0.12[†]	0.21[§]
	Affiliation		0.26[§]
Men (N=213)	Rank	−0.03	0.21[‡]
	Affiliation		0.27[§]
Women (N=75)	Rank	−0.27[†]	0.20[*]
	Affiliation		0.24[†]

NOTES: The variables *rank, affiliation,* and *productivity* are residuals of regressions of the original variables on type of fellowship, fields, and years since doctorate. The correlations are partial correlations between two variables, controlling for the third. The original variables were productivity: number of publications per years since doctorate; rank: 1 = nonprofessorial positions, 2 = assistant professor, 3 = associate professor, 4 = full professor; affiliation: 0 = not at top 15 percent of graduate departments, 1 = at top 15 percent of graduate departments.
See table 2.2 for explanation of reference marks in this table.

ity was found to have some effect on the prestige of institutional affiliation (Allison and Long 1987), the reverse effect—institutional affiliation impacting productivity—was found to outweigh it (Allison and Long 1990; Long 1978; Long and McGinnis 1981).

We should repeat here that academic rank belongs to the social system of academe rather than to the social system of science. Technically, rank promotion is a reward in the former, not in the latter, system. To the extent, however, that personnel decisions in the social system of academe are made on the basis of scientific values, rank becomes a reward for scientific performance. And, in practice, scientific merit forms a central component of academic merit, especially at research universities.

Affiliation and Productivity

Using residuals for affiliation and productivity that controlled for type of fellowship, fields, and years since doctorate, the partial correlation between affiliation and productivity, net of rank, was $r = 0.26$

($p = 0.0001$). The men's correlation was $r = 0.27$ ($p = 0.0001$), and the women's was $r = 0.24$ ($p = 0.0395$) (table 2.3). Thus, affiliation and productivity were correlated in our sample, both for the men and for the women. Regardless of the exact nature of the causal relationship between location and productivity, it appeared to be similar for men and women.

Rank and Productivity

The partial correlation between the residuals for rank and productivity, net of affiliation, was $r = 0.21$ ($p = 0.0003$) for the whole sample, $r = .21$ ($p = 0.0019$) for the men, and $r = 0.20$ ($p = 0.0928$) for the women. Thus, productivity was correlated with rank in similar ways for men and women. The strength of the correlation of productivity with rank was comparable to that of the correlation of productivity with affiliation.

Rank and Affiliation

The already discussed trade-off relationship for women as a group between high academic rank and prestigious institutional affiliation can be discerned in a negative partial correlation for women between rank and affiliation ($r = -0.27$, $p = 0.0212$). By contrast, men as a group did not experience a significant tradeoff ($r = -0.03$, $p = 0.6486$).

Rank, Affiliation, and Productivity

We now consider all three outcome variables together, using two as controls in regressions of the third. Controlling for rank and institutional affiliation, do women produce fewer publications on the average? And, controlling for publication productivity and institutional affiliation, are women scientists at a rank disadvantage compared with men?[5]

With respect to the first question, recall that the raw gender difference in productivity was small but significant. Controlling for

Table 2.4. Regression Models of Productivity on Years Since Doctorate, Fellowship, Fields, Affiliation, Rank, and Gender

	1	2
Years since doctorate	−0.02	−0.03
Fellowship	0.07	1.47
Biology	0.90[†]	0.29
PSME	0.97[‡]	0.39
Affiliation	1.09[§]	1.55[†]
Rank	0.49[§]	0.84[*]
Gender	−0.27	0.80
Gender × years since doctorate		0.02
Gender × fellowship		−0.98
Gender × biology		0.48
Gender × PSME		0.39
Gender × affiliation		−0.38
Gender × rank		−0.31
R^2	0.1397	0.1573
N	288	288

NOTES: Unstandardized regression coefficients. Productivity: number of publications per years since doctorate. Rank: 1 = nonprofessorial positions; 2 = assistant professor; 3 = associate professor; 4 = full professor. Affiliation: 0 = not at top 15 percent of graduate departments; 1 = at top 15 percent graduate departments. Fellowship: 0 = NRC; 1 = NSF. Gender: 1 = male; 2 = female. See table 2.2 for explanations of reference marks in this table.

variables such as fields, fellowship, and years since doctorate weakened this gender difference. Additionally, rank can account for a part of the gender difference because women hold lower ranks on the average and rank predicts productivity. Controlling for rank and affiliation in addition to the regular control variables, women respondents were estimated annually to publish a quarter of a publication (0.27) less than men did, but this difference was too small to be statistically significant ($p = 0.2981$) (table 2.4, model 1). There were no significant gender interactions (table 2.4, model 2).

The second question asked about gender differences in rank achievement, after controlling for publication productivity and affiliation. In a main-effects model (Table 2.5, model 1), including among the predictors productivity and institutional affiliation in addition to

Table 2.5. Regression Models of Rank on Years Since Doctorate, Fellowship, Fields, Affiliation, Productivity, and Gender

	1	2	3
Years since doctorate	0.08§	0.09§	0.09§
Fellowship	0.34‡	1.01‡	1.02‡
Biology	−0.35‡	−1.24‡	−1.34†
PSME	−0.22	0.46	0.42
Affiliation	−0.20†	0.50*	0.34
Productivity	0.09§	0.07	0.07
Gender	−0.26†	0.24	0.16
Gender × years since doctorate		0.00	0.00
Gender × fellowship		−0.48*	−0.49*
Gender × biology		0.62†	0.72*
Gender × PSME		−0.58*	−0.51
Gender × affiliation		−0.58†	−0.38
Gender × productivity		0.00	0.00
Affiliation × biology			0.39
Affiliation × PSME			0.15
Gender × affiliation × biology			−0.33
Gender × affiliation × PSME			−0.24
R^2	0.4865	0.5502	0.5515
N	288	288	288

NOTES: Unstandardized regression coefficients. Rank: 1 = nonprofessorial positions; 2 = assistant professor; 3 = associate professor; 4 = full professor. Fellowship: 0 = NRC; 1 = NSF. Affiliation: 0 = not at top 15 percent of graduate departments; 1 = at top 15 percent of graduate departments. Productivity: number of publications per years since doctorate. Gender: 1 = male; 2 = female.
See table 2.2 for explanations of reference marks in this table.

the usual controls, women's expected rank is a quarter lower than that of comparable men (p = 0.0165). Thus, women's lower rank achievement, compared with men, cannot be explained by their lower productivity because women's estimated rank is lower at each productivity level.

Considering a model with gender interactions (Table 2.5, model 2), it again becomes clear that the women's rank disadvantage was concentrated in the fields outside biology; among the biologists, however, women's rank was similar to men's rank on the average. Thus, the threshold pattern in biology contrasted with the glass-ceiling pattern in the other sciences. There was also a trade-off between affiliation and rank for women as a group but not for men ($p = 0.0106$). To examine whether this tradeoff also existed in biology (where, as we have mentioned, women as a group did not experience any disparities in terms of rank) three-way interactions of fields, affiliation, and gender were included in the regression model (table 2.5, model 3). These three-way interactions failed to be significant; but in regressions for biologists alone, there was no significant trade-off, which indicated that it may be primarily a problem for women scientists outside biology.

In sum, scholarly merit fell short of explaining the observed gender disparities in the key outcome of academic rank. Does this mean that the disparities were caused by gender discrimination? Before making such an inference, the sophisticated-residualist approach suggests that we should examine alternative explanations. Hence, we next turn to human-capital variables in the respondents' past, going back to personal background, graduate school, and postdoctoral characteristics. This will also allow us to determine gender irregularities in the accumulation of advantages and disadvantages in early segments of the science pipeline.

3

ACCUMULATING HUMAN CAPITAL: STAGES IN THE SCIENCE PIPELINE

WOMEN "ALWAYS HAD TO be better in order to be equal." In these or similar words, several of our respondents echoed a widespread notion about the conditions of women's success in science. In this chapter, we examine part of this common perception by turning back to our respondents' earlier career stages and examining whether the women in our sample, as a group, had certain advantages that helped them succeed in an environment that was relatively less welcoming toward women.

The adage at the beginning of this chapter could be interpreted as an informal statement of the compensation model. Those relatively few women who succeeded in their science career paths to the point of winning a prestigious postdoctoral fellowship may have had an especially bountiful amount of human capital that helped them compensate for the obstacles and disadvantages typically awaiting women on their path toward becoming scientists. Thus, a compensation model was a guiding hypothesis in our examination of gender differences in these early stages. In terms of personal background variables, for instance, our reasoning went as follows. Because being a female child already constitutes a certain handicap against pursuing a career in most sciences, other favorable circumstances or characteristics in our female scientists' personal background may have had to compensate for this handicap. Therefore, we expected such favorable circumstances or characteristics to be more prevalent among our women respondents than our men respondents.

Within this framework, we should stress the distinction between early gender differences and gender disadvantages in terms of later career outcomes. Gender differences are not automatically gender disadvantages. Nevertheless, the theory of the accumulation of advantages and disadvantages emphasizes that an individual's characteristics at early stages may to some extent determine later career outcomes. Thus, some gender differences at early stages may translate into gender disadvantages if they are associated with later outcomes.

We distinguished the following types of early characteristics that are associated with later outcome. (1) If an early characteristic on which the genders do not differ has a main effect on outcome, it constitutes a gender-neutral advantage or disadvantage. (2) A gender interaction on such a characteristic, however, indicates a gender-specific accumulation of advantages or disadvantages. In the case of an early characteristic on which the genders differ, both (3) a main effect and (4) a gender interaction indicate gender-specific accumulation. In the last case, the gender interaction may either magnify or attenuate the initial difference, depending on its direction. In the last category are the "messiest" cases of a gender-specific accumulation of advantages and disadvantages that obfuscate the degree of universalism or meritocracy in the science system and thus make it hard to pinpoint "process" gender discrimination. An example of a gender-specific accumulation of disadvantage that magnifies a previous disadvantage for women would be an early characteristic that fewer women than men share and that is then associated with considerably higher rank for men but with only marginally higher rank for women. In this case, women as a group would face two kinds of disadvantages. First, fewer women than men would have that advantageous characteristic; and even those women who do have it would not benefit from it, on the average, as much as men do.

In this chapter, we glance into our respondents' past—surveying their characteristics and experiences at various early stages along the science pipeline. We start with the former fellows' personal background—including, for instance, their parents' educational level. Next, the respondents' graduate school phase is examined, followed by a review of their postdoctoral experience.

Personal Background

The following summary of the personal background characteristics of our respondents will particularly focus on gender differences. The areas covered range from sociological characteristics (educational background of parents) and social psychological characteristics (sibship size, birth rank, and siblings' gender) to psychodynamic experiences (early illnesses and losses). As noted earlier, a compensation model guided our thinking about gender differences in personal background variables. In the following, we denote what circumstances can be considered favorable in the areas of parental education, sibship position, and early illnesses and losses and then examine whether our sample indeed contains gender differences of the predicted kind.

Parents' Education

We anticipated that a high-parental education level would counteract the pressures that generally discourage girls from aspirations in the science field. First, the parents of the women scientists were expected to have a higher educational level, on the average, than the parents of men scientists. Second, a highly educated mother may be a particularly important source of encouragement and support for females to enter a science career. Thus, we expected the difference in educational level between mothers of women scientists and mothers of men scientists to be larger than the difference in the educational level between fathers of women scientists and fathers of men scientists. Third, we expected a convergence of the parents' educational background of women and men scientists over the years. As the cultural acceptance of, as well as institutional support for, women scientists has grown, more women from less-educated backgrounds may have succeeded in pursuing a science career.

These hypotheses tested well in general. Neither parent of 42.8 percent of the men respondents had any college degree, whereas this was the case for only 29.8 percent of the women. Both parents had at least bachelor's degrees for 31.7 percent of the men and 41.0 percent

Table 3.1. Educational Level of Parents (in Percent)

	Father		Mother	
	Men	*Women*	*Men*	*Women*
Less than completed high school	15.7	13.7	12.3	7.3
High school diploma	20.9	18.1	34.6	29.2
Some college	10.3	5.0	10.3	7.9
Associate's degree	2.6	3.3	4.4	3.4
Bachelor's degree	17.3	20.9	21.1	33.2
Some graduate school	4.6	2.8	3.6	2.8
Master's or professional degree	12.9	20.3	10.5	12.4
Some post-master's work	1.6	0.6	1.0	2.8
Doctoral degree	14.1	15.4	2.2	1.1
N	503	182	503	178

of the women. The mean educational level for both parents was clearly higher for women (4.4) than for men (4.0) on a nine-point scale of educational levels ($p = 0.0286$).[1]

The mothers of women respondents were, on the average, more highly educated than the men's mothers (table 3.1). More than half (52.3 percent) of the mothers of our women respondents had at least a bachelor's degree as compared with only 38.4 of the mothers of our men respondents. The mean educational level of the mothers of women respondents was 4.04, which was about half a level higher than the mean educational level of the mothers of men respondents (3.58) ($p = 0.0115$). Educated mothers may have been particularly keen to impart the value of scientific endeavor to their daughters or at least not to downgrade such ambitions.

Similar differences were found for the educational level of the respondents' fathers, but the difference in educational level was not quite as pronounced. Of the female respondents' fathers, 59.9 percent had at least a bachelor's degree, whereas for the male respondents' fathers the figure was 50.5 percent. The average educational level was 4.84 for the fathers of women respondents and 4.43 for the fathers of men respondents ($p = 0.0824$). Although the gap between women's and men's mothers' educational level was, as expected, slightly larger than the gap between women's and men's fathers' educational level, this difference did not reach significance (repeated measures analysis of variance; $p = 0.6069$).

A special look is warranted for the older respondents in our study, who grew up in a time when it was more difficult for women to become scientists than it has been in recent years. Among the older respondents (those who received their doctorate before 1978), the gender difference in the mothers' educational level was substantial. The women's mothers averaged 4.09 as compared with 3.31 for mens' mothers. Among the younger respondents (who received their doctorate in 1978 or after), the average education level of men's mothers (3.92) had almost reached that of women's mothers (4.01). Among the older respondents, there was also a large gender gap in fathers' educational background (women's fathers: 5.17; men's fathers 4.06). For the younger group, however, the men's fathers' average educational level (4.88) had even slightly surpassed the women's fathers' educational level (4.57).

In short, the parental educational level differed for our female and male respondents in roughly the expected ways. Our women respondents had, on the average, more highly educated mothers than our male respondents did. The women's fathers were also more highly educated, but differences in fathers' educational level were slightly less pronounced. Moreover, the educational backgrounds of the parents of women and men respondents tended to converge over time. Among the older scientists, women's parents were much more highly educated than men's parents, on the average. Whereas the educational level of the men's parents increased in time—probably in line with an increase in the formal level of education among the general population—the women's mothers' education stayed the same, and the women's fathers' education even dropped. Thus, the number of women former postdoctoral fellows from less-educated backgrounds has grown as the idea of a woman scientist has become more normal.

Siblings

Ever since Galton's (1875) classic study, researchers have investigated the effect of birth order and sibship size on an individual's traits, particularly on cognitive achievement (Adams 1972; Schooler 1972; Steelman 1985) or even on scientists' predisposition to espouse revolutionary ideas (Sulloway 1989). In recent years, Zajonc's (1976) work has received wide, and sometimes critical attention (for example, Retherford and Sewell 1991). Zajonc found that, within each sibship size, firstborns received the highest scores on intelligence tests and that intelligence, so measured, declined with later birth rank. Intelligence for each birth rank, was also found to decline with increasing sibship size. The exception was that single children were less intelligent than the firstborns in small sibships. Other researchers have pointed out that firstborns typically receive more attention from their parents and internalize parental expectations to a higher degree than later borns; therefore, they develop more elevated standards of achievement (Kagan 1971, 148–150).

A higher level of intelligence and an overachieving outlook on life among firstborns led us to expect that they are overrepresented among scientists. That, indeed, has been repeatedly found. Francis Galton noted an overrepresentation of firstborns among the fellows in the Royal Society, and firstborns were also overrepresented among the scientists listed in the "American Men of Science" compilation (Visher 1947) as well as among Nobel Prize winners (Clark and Rice 1982).

The association of birth order and sibship size with a career in science has usually been examined for scientists in general (which means mostly for male scientists), but we focused on possible gender differences in this association. We again hypothesized a compensation effect. Favorable family constellations were expected to be more prevalent among our sample's women to compensate for the cultural and social pressures that impede women's aspirations for a science career. We expected that, in our sample, women scientists came from even smaller sibship sizes than men scientists and that firstborns were even more strongly overrepresented among women scientists than among men scientists.

In addition, researchers have posited that the gender composition of sibships is important, but they disagree about its specific effect. The two antithetical theories can be called the surrogate-son theory and the brother-impact theory. The first theory says that the absence of sons may turn daughters into surrogate sons onto whom fathers project their ideas about a career suitable for a son (Weitzman 1984, 218). Moreover, in all-female groups of siblings, as some of our female interviewees emphasized, girls perform activities that are not considered typically female and possibly would have been taken over by the boys if there had been any. In her male sample of eminent scientists, Roe (1952) noted the absence of older brothers among the facilitators of scientific success. This pattern, one might hypothesize, applies to women scientists to an even greater extent because the older brother's role as the crown prince may be even more pronounced vis-à-vis a younger sister than a younger brother. Following these approaches, one would expect an overrepresentation of women without brothers among scientists.

The second theory, instead of emphasizing the effects of parental socialization, stresses the importance of socialization among the siblings. In this view, brothers, and especially older brothers, encourage and reinforce male behavior patterns in their sisters. For instance, girls with older brothers were found more likely than girls without them to exhibit typically male traits (Brim 1958). This suggests a facilitating role among older brothers of female scientists and consequently an increased proportion of women scientists with older brothers—or perhaps with brothers in general.

In short, we examined possible gender differences in three aspects of our respondents' sibship constellations: size, birth order, and gender composition. Contrary to the first hypothesis, men (with an average sibship size of 3.03 siblings) and women (3.18 siblings) came from families of roughly the same size. The distribution of sibship sizes was remarkably similar for both genders. Of our respondents, 14.4 percent were only children (men: 14.7 percent; women: 13.6 percent); 28.5 percent (men: 29.0 percent; women: 27.2 percent) came from two-sibling and 26.5 percent (men: 25.6 percent; women: 28.8 percent) from three-sibling families.

Second, in terms of birth order, firstborns were indeed overrepresented among our respondents but similarly among both genders. Given the sibship sizes of our respondents, we would expect 205 (34.5 percent) firstborns among the 595 scientists with at least one sibling if firstborns occurred randomly in our sample.[2] In reality, there were 292 (49.1 percent) firstborns ($p < 0.0005$). This overrepresentation of firstborns occurred similarly among men (48.4 percent actual versus 34.7 percent expected) and women (50.9 percent actual versus 33.9 percent expected).

Third, we looked at the gender composition of our respondents' sibships in general as well at the existence of *older* brothers only. In neither respect did we find substantial deviations from chance. In addition, there was not much difference between the genders.

Within this overall picture of gender similarity, there were two hints of gender differences, both of them in two-sibling families. Among the men respondents with one sibling, firstborns (51.4 percent) and secondborns (48.6 percent) occurred about equally; but almost two-thirds (63.5 percent) of the women were firstborns. Although this gender difference in the proportion of firstborns was not statistically significant (LRCS = 2.287; p = 0.130), it was in the predicted direction and might indicate that in two-sibling families, secondborn women face somewhat greater impediments against a science career than secondborn men do. For females, the handicap of being later born may be a factor in the two-sibling family, whereas for males it may only become effective in sibship sizes of three and larger. It seemed as if parents tended to give equal attention to secondborn boys' scientific development, but secondborn girls might have been less able to attract such attention or ambition.

The other difference in two-sibling families was that women were less likely than men to have a sister (38.5 percent versus 56.2 percent; LRCS = 4.837, p = 0.028). If anything, this would support the brother-impact theory over the surrogate-son theory. One can speculate that the presence of a brother may, on the one hand, encourage general behavior patterns in girls (such as competitiveness and aggressiveness) that are useful for later science careers. On the other hand, scientifically inclined brothers may have an even more direct impact—introducing their sisters to science activities, even if those have a male image.

Early Losses and Physical Handicaps

Roe (1952) conducted one of the early studies of high-achieving scientists' personality traits. Using a psychodynamic approach, Roe focused on childhood experiences. She noted that scientists were particularly self-sufficient, self-contained, and nonsociable, even in their early lives. As one of the causes for their turning away from social interaction, Roe identified the childhood loss of a parent or a severe illness. Roe's findings were based on an all-male sample of scientists. Our interest was to see whether female and male scientists in our sample differed in the experience of early losses and handicaps. Again we would expect a higher incidence of losses and illnesses in women to compensate for their cultural handicap. Information about these issues was available only for former NSF postdoctoral fellows.

Our data indicated that by the time they reached the age of twenty-one, very few former NSF fellows had been seriously ill (2.6 percent of the men; 1.0 percent of the women) or were physically handicapped (men: 0.3 percent; women: 0.0 percent). As to traumatic disruptions of the family during childhood, no significant gender difference showed up in terms of father's death by the respondent's age of eighteen (men: 5.5 percent; women: 4.0 percent) or parents' divorce or separation by the same age (men: 5.0 percent; women: 7.0 percent).

A striking gender difference, however, was found in the incidence of the loss of a mother. A much higher fraction of women (5.0 percent) than men (0.8 percent) had, at an early age, experienced the death of their mothers (LRCS = 6.304; p = 0.012).[3] In light of the small numbers involved, this finding may, of course, be little more than a statistical fluke. But we hesitate to dismiss it because this is precisely what a combination of Chodorow's (1974) and Roe's (1952) theories would predict—that the mother's death is in some ways a stronger stimulus to enter science for girls than for boys.

As we mentioned earlier, Chodorow (1974) argued that women's greater connectedness with others and men's greater independence result from the fact that young girls remain more closely connected with their mothers, whereas individuation and separation from the mother are key processes in young boys' development. For girls, the mother's death could thus be considered a particularly grave trauma that forces separation and individuation upon them. A woman interviewee, who was five years old when her mother died and six years old when a somewhat hostile stepmother arrived, pointed out how these events prevented her from forming close relationships with other people. "I didn't really have any real close human relationships until I met my husband, I would say, and had my children, because my mother died when I was five, and my father was kind of a distant person." According to Roe, such difficulties with social relationships tend to precipitate a turn to the objective world of scientific investigation. This psychodynamic mechanism linking mother's death to daughters' science careers is, of course, highly speculative; and there may be a number of possible explanations for why the mother's death may tend occasionally to lead a daughter toward a career in science. The really intriguing question may be, after all, why so few sons whose mothers died opted for a science career.

To summarize, as we expected according to the compensation concept, the parents of women former postdoctoral fellows were more highly educated than the men's parents. This gender difference appeared to decrease over the years. Little evidence for a compensation mechanism was found in the area of sibship constellations, which, with a few exceptions, were remarkably similar for women and men. Some indication of a compensation effect existed in the area of traumatic childhood events. The most striking gender difference in this area was that women respondents were much more likely than men to report that their mothers had died by the respondents' age eighteen.

Personal Background, Early Losses, and Career Outcomes

Our respondents were similar in their achievement at a relatively late stage in the science pipeline—receiving a prestigious postdoctoral fellowship from the NSF or NRC. We therefore expected that early personal background variables would not strongly affect career outcomes, and overall this was the case. Nonetheless, some effects could be discerned in multiple regressions (or multiple logistic regressions) that included the control variables of type of postdoctoral fellowship, year of doctorate, and fields.

If small sibship size is associated with tenacity, a strong intrinsic work motivation, and possibly a lack of sociability, it is not surprising that academic scientists had fewer siblings on average ($p = 0.0291$) compared with those who work in nonacademic science. Similarly, it can be explained that women from small sibships were less likely than women from large sibships to leave research science. It is much more puzzling, however, that the opposite trend applied to men ($p =$

0.0394 for interaction). One can speculate that, whereas women showed the tenacity in their science careers that appears typically associated with smaller sibship sizes, their male counterparts were more likely to pursue other career options that may not have been so available to women.

Among the current academic scientists, high rank (controlled for institutional prestige) correlated with large sibship size ($p = 0.0378$). It may be enlightening to see this result in combination with the two nonsignificant trends that academic scientists from smaller sibships tended to produce more publications ($p = 0.1110$) and firstborns were likewise more productive than later borns ($p = 0.1130$). Thus, respondents from large sibships were more successful in terms of academic rank even though they were marginally less productive in terms of publications. An explanation that is consistent with theories about sibship size would point out that growing up in large families may stimulate the acquisition of social and political skills. By contrast, people from small families may be extremely industrious and hard-working (producing a large amount of publications) but may lack some of the social skills conducive to rank advancement.

Academic scientists were more likely to have no brothers than the scientists currently working outside academe ($p = 0.0149$). Among academic scientists, men without older brothers were slightly more successful in terms of rank than those with older brothers. This finding for males echoes Roe's (1952) finding that the absence of an older brother facilitates scientific success. In contrast, women *with* older brothers were markedly more successful in terms of academic rank than women with no older brothers ($p = 0.0412$ for gender interaction). This particular result supports the brother-impact theory over the surrogate-son theory for females. It suggests that Roe's (1952) finding was specific to male scientists only and that the situation is different for females.

In terms of parents' educational level, there were no significant effects but weak trends in the expected direction. Although a highly educated background was somewhat associated with persistence in science and high academic rank, it did not seem to matter much in our group of former postdoctoral fellows.

The numbers of NSF respondents who had experienced a severe illness or physical handicap were too small to investigate possible effects of these experiences on career outcomes. Because the incidence of any particular type of family disruption was also low, we formed a composite variable aggregating various kinds of family disruption (death of either parent or parents' divorce or separation before respondent's age eighteen). This applied to 11.5 percent of our NSF respondents (men: 10.8 percent; women: 14.0 percent). Compared with those from nondisrupted families, respondents from disrupted families were more likely to leave science altogether ($p = 0.0565$) and to work in non-academic science rather than in academic science ($p = 0.0679$); but those who had gone on to academic careers were more productive in terms of publication output ($p = 0.0876$)— possibly because the less-productive ones had already left. Thus, if there were any long-term effects of family disruption, they polarized the outcomes between becoming an unusually productive academic scientist or not working in academic science at all.

Graduate School

Graduate school is a crucial phase in a scientist's career. During graduate school, students are socialized into the role of being a scientist, which includes both formal skills and requirements for doing research and informal skills of behaving and acting like a scientist.

This section surveys the graduate-school experience of our respondents, focusing on gender differences. We would like to emphasize again that our sample does not represent the general population of American graduate students but consists of an elite subgroup. Their characteristics as a group, as well as gender differences among them, are expected to differ from the general population in the way the compensation model would predict. Thus, we expect the women in our sample to have had more advantages than women graduate students in general. Even if, for instance, in the general population of graduate students, fewer women than men have certain positive characteristics (such as graduate fellowships), the relatively few women selected for the prestigious postdoctoral fellowships should be more likely to have such characteristics than the average woman graduate student and should thus resemble their male cohorts to a higher degree.

In the following two sections, we present gender differences and gender disadvantages in a chronological and systematic way (first for the graduate experience, then for the postdoctoral experience). Then we look at graduate and postdoctoral experiences in conjunction, identifying some major themes and interpreting them within our theoretical framework of the accumulation of advantages and disadvantages.

Gender Differences in Graduate School Characteristics

To examine whether women respondents were at a disadvantage in terms of how graduate school prepared them for their future careers, we examined a number of graduate characteristics that were identified by various researchers as having an impact on later career development or that appeared likely to have such an impact.

First, there were a few general timing parameters of graduate study. We looked at the time elapsed between bachelor's and doctorate degrees (in years), at the actual time spent in graduate school (in years), at the incidence of part-time study, interruption, and at a variable combining part-time study and interruption.[4]

The next group of variables described the graduate students' social integration into the department and research community. These variables indicated whether the scientists held a research assistantship during graduate school, the academic rank as well as the tenure status of the supervisor, and whether the supervisor held a position of particular influence in the department. The supervisor's gender was also noted. Another variable recorded whether the students worked most closely with their advisors or with other scientists; and the next variable was a scale indicating the degree of collaborative research style—ranging from 1 ("usually worked alone on my scientific problems") to 5 ("my research was performed almost exclusively in collaboration with other scientists").

We also wondered how the respondents financed graduate school. Variables in this area indicated whether the respondents incurred any educational debts, whether they held graduate fellow-

Table 3.2. Means of Graduate Characteristics by Gender

Variable	Men	Women
Time in graduate school		
Time elapsed between degrees	5.72	7.01§
Study time elapsed between degrees	5.00	5.38‡
Part-time study	0.112	0.241§
Interruption	0.220	0.435§
Part-time study or interruption	0.275	0.497§
Social integration		
Research assistantship	0.731	0.681
Advisor's rank	3.547	3.365‡
Advisor's special status	0.283	0.288
Advisor's gender	1.026	1.079†
Advisor's tenure status	0.867	0.835
Close collaboration with advisor	0.774	0.720
Collaborative research style	2.10	2.17
Financing graduate school		
Debt	0.301	0.328
Fellowship	0.758	0.691*
Nonuniversity funds (%)	0.174	0.183
Family status		
Married	0.559	0.563
Parent	0.246	0.178*
Training and performance		
Top university	0.541	0.382§
Publications	0.729	0.753
Number of publications	3.61	2.89†

NOTE: See table 2.2 for explanations of reference marks in this table.

ships, and to what proportion their graduate studies were funded by nonuniversity sources (such as unrelated jobs or family support).

In terms of training and performance, we noted whether the respondents attended a department rated among the top 15 percent of graduate schools, whether they had any predoctoral publications, and the number of those predoctoral publications.

The last group of variables focused on the graduate students' personal status. We inquired if the respondents were married and whether they were parents in graduate school. Table 3.2 displays these graduate-school characteristics by gender.

Before discussing in more detail the characteristics on which the genders differ, we should note the high overall degree of gender similarity among our respondents. This is indeed what we expected according to the compensation model. For instance, whereas national statistics have shown that fewer women than men graduate students held research assistantships (NSF 1986, 27), such a difference is hardly noticeable in our group. And for the total population of graduate students in the sciences, it has been stated that "sizeable sex differences occur in amount and kind of financial aid, and all of them are disadvantageous to women" (Chamberlain 1988, 210). In our sample, however, the sources of financial support were remarkably similar for men and women as indicated by the proportion of graduate-school funding that came from nonuniversity sources as well as by the proportion of respondents with educational debts. The possible exception was that a smaller number of women than men held graduate fellowships. After controlling for fields, the years when graduate school was attended, and the type of postdoctoral fellowship (NSF or NRC), we noted that this difference also became nonsignificant. Similarly, the differences in advisor rank and gender as well as in the graduate department prestige faded once the control variables were considered.

Few gender differences remained. Although about the same proportion of women and men had at least one predoctoral publication, the average number of publications was lower for women than for men. Fewer women were parents during graduate school. The most substantial differences were in the general *smoothness* of the graduate phase. Women were more likely than men, on the average, to interrupt their studies or study part time. They spent a longer time between bachelor's degree and doctorate in terms of both total time elapsed and actual study time.

Time in Graduate School

It has repeatedly been found that a speedy completion of graduate study presages later success as a scientist (Clemente 1973; Hagstrom

1971). Women have generally been observed to take longer than men to receive their doctorate, on the average. Among all persons who received their doctorate in science or engineering between 1960 and 1986, men spent a median of 7.3 years between baccalaureate and doctorate, whereas women spent 8.0 years (NSF 1988, 139; see also Centra 1974). This gender gap, however, has apparently been shrinking over the years (NAS 1979).

Compared with the national statistics, our respondents achieved their doctorate somewhat faster (men: 5.7 years; women: 7.0 years). In a marked gender difference, women respondents spent an average of 1.3 years more between bachelor's and doctorate degrees than men did ($p = 0.0001$).

We now asked to what degree this gender difference resulted from the women's spending a longer time studying and to what degree the difference could be explained by a higher incidence of substantial interruptions or of part-time study among women. Women were more likely than men both to interrupt their studies (43.5 percent versus 22.0 percent; $p = 0.0001$) and to study part time (24.1 percent versus 11.2 percent; $p = 0.0007$). Only half the women (50.3 percent) but almost three-quarters of the men (72.6 percent) were full-time students all the time between their bachelor's and doctorate degrees ($p = 0.0001$). Interruptions were more frequent among younger than older respondents (31.5 percent versus 24.7 percent; $p = 0.0016$), as well as among biologists (31.5 percent) and social scientists (45.7 percent) compared to PSME scientists (22.0 percent; $p = 0.0017$).

After controlling for interruptions and part-time study by counting only the actual study time, we noted that the gender difference in time spent between bachelor's degree and doctorate dropped from 1.3 years to 0.4 years (men: 5.0 years; women 5.4 years; $p = 0.0029$). The women's higher propensity for interruptions and part-time study accounted for more than two-thirds (70.5 percent) of the initial gender difference in time spent between bachelor's and doctorate degrees, but women still spent a significantly longer time on their actual graduate studies. After controlling for fields, year of doctorate, and type of postdoctoral fellowship (NSF or NRC), however, we noted that the gender difference in the actual duration of studies between bachelor's degree and doctorate disappeared.

In sum, the straight and linear progression from college to full-time graduate school to doctorate was much more typical for our male than for our female respondents. There is nothing mysterious about our women respondents' longer graduate phase. One need not speculate, for instance, about a lack of drive or talent among the female respondents. Women's longer average interval between bachelor's and doctorate degrees can be explained by their choice of field and, most of all, the higher frequency of part-time study and interruptions. We now wondered how these interruptions were related to family responsibilities.

Parental Status

Fewer women than men respondents were parents during their time in graduate school (men: 24.6 percent; women: 17.8 percent; $p = 0.0349$). We hypothesized that children prolonged the time in graduate school, particularly for women. Put in more technical terms, we expected an interaction between gender and children during graduate school. Such an interaction was indeed found ($p = 0.0001$). Whereas children prolonged the men respondents' time by an estimated 0.9 years, they prolonged the women's time between bachelor's and doctorate by 3.1 years. Note that for the actual time spent in graduate school (that is, apart from leaves or part-time study), having children made no difference ($p = 0.9408$) for men or women (interaction: $p = 0.6657$).

As to the likelihood of being a part-time student and of interrupting, we again expected interaction effects between gender and having children (having children should increase the probability of being a part-time student or of interrupting for women more than for men). Both interactions were significant (part time: $p = 0.0016$; interrupting: $p = 0.0073$).

We expected that the gender differences in the effects of having children on time between bachelor's and doctorate would decline with time, in step with a general weakening of the traditional gender division of labor in the past two decades. The genders should become more similar, but the question is, Does the prolonging effect of parenthood shrink for women or grow for men? The three-way interaction of gender, year of doctorate and children was indeed significant in predicting the time period between the degrees ($p = 0.0165$). This three-way interaction is easier to interpret when regressions are considered separately for men and women. For the younger respondents among the men, parenthood extended the predicted time spent between bachelor's degree and doctorate much longer than for older men ($p = 0.0118$). A man who graduated in 1960 spent about the same time, on average, in graduate school regardless of whether he had children or not. By contrast, for a man who graduated in 1980, being a parent prolonged the time between bachelor's and doctorate degrees by an estimated 1.6 years. For women, on the other hand, the prolonging effect of parenthood on the period between bachelor's degree and doctorate decreased with time, albeit not significantly ($p = 0.4523$). Thus, the gender convergence was due to movements from both sides, although change was statistically better measurable for the men.

Outcomes

We now examine whether any graduate-student characteristics or experiences were related to later career outcomes and whether these relationships differed by gender. Again, we should stress that we are dealing with a restricted range of subjects—students who were exceptionally successful in graduate school. Moreover, the common characteristic of winning and going through a prestigious postdoctoral fellowship may have weakened, if not obliterated, the effects of any prior differences in graduate school. Thus, we do not expect strong effects of graduate characteristics on career outcomes, but those that do exist are so much the more interesting.

In terms of career outcomes, we focus first on whether the former postdoctoral fellows have subsequently left science; second, on whether the respondents who are still scientists are employed in the academic or nonacademic field; and third, on those respondents currently in academic science. Here the outcome variables are academic rank, controlled for institutional prestige, as well as publication productivity. All these outcome variables will be controlled for academic age (years since obtaining doctorate), academic fields, and type of fellowship (NSF or NRC).

A regression approach was used to determine if any graduate-study characteristics had a statistically significant association with the respective outcome variables and then if there were any gender differences. For dichotomous outcome variables, logistic regressions were employed. Two major difficulties were that the number of predictor variables was high compared with our limited sample size, and that some of the predictors strongly correlated with each other. To deal with these problems, we conducted regressions (or logistic regressions) with a smaller number of principal components and also carried out stepwise regressions (or stepwise logistic regressions), which reduced the model to a small group of significant variables.[5]

The small sample size, and especially the relatively small number of women, limited us in documenting the women's unique experience. In this book we present the coefficients for the whole sample of postdoctoral fellows and then determine whether the women's situation was significantly different from the men's (that is, whether there were significant gender interactions). A more complete treatment, feasible with larger samples, would have presented full regression models for men and women separately, determined the significance of the coefficients separately, and then tested whether the coefficients differed by gender. Therefore, the gender differences reported here should not be regarded as an exhaustive list but rather as the strongest few.

Graduate-study Characteristics and Later Leaving Science

The strongest graduate study predictor of later leaving science was having had an interruption of graduate study between bachelor's and doctorate degrees (table 3.3). Smoothness of the graduate-school phase may indicate great internal commitment to a science career as well as the absence of external problems that impede being a professional scientist. Similarly, men who studied part time at some point between their bachelor's and doctorate degrees were more likely to leave science later on. The opposite, however, was true for women. Women respondents with a period of part-time study

83

Table 3.3. Graduate-study Predictors of Leaving Science

	Full model	Model with significant variables	Interaction model
Controls			
Fellowship	0.16	−0.09	0.17
Academic age	0.06‡	0.05‡	0.05‡
Biology	−0.42	−0.20	−0.40
PSME	−0.83*	−0.65	−0.72*
Time in graduate school			
Time elapsed between degrees	−0.06		
Study time elapsed	0.02		
Part-time study	0.19		2.10*
Interruption	2.32†	1.06§	0.99‡
Part-time study or interruption	−1.32		
Social integration			
Research assistantship	0.35		−0.00
Advisor's rank	−0.01		
Advisor's special status	−0.49		
Advisor's gender	−0.41		
Advisor's tenure status	0.21		
Close collaboration with advisor	0.78*	0.80†	0.73*
Collaborative research style	−0.20		
Financing graduate school			
Debt	0.26		
Fellowship	−0.18		
Nonuniversity funds (%)	0.00		
Family status			
Married	−0.09		
Parent	0.14		
Training and performance			
Top university	−0.44		
Publications	0.08		
Number of publications	−0.00		
Gender interactions			
Gender			0.13
Gender × part-time study			−1.74*
Gender × research assistantship			0.20
N	643	643	677
MCS	37.8	27.1	31.7
Pseudo-R^2	0.0982	0.0704	0.0765

NOTES: MCS (model chi-square): reduction in chi-square compared with the chi-square of a baseline model containing only the intercept when the model variables are added. Pseudo-R^2 = [MCS]/[baseline chi-square].
See table 2.2 for explanations of reference marks in this table.

between baccalaureate and doctorate were *less* likely to leave science than those without part-time study.

The men's pattern is what one would expect. A speedy completion of the doctorate has been interpreted as an indication of strong ability or motivation (Folger, Astin, and Bayer 1970, 265). In this interpretation, a graduate-school phase that is prolonged through part-time study would indicate a lack of commitment or competence. Our results suggest that such an assumption may be only warranted for men and does not do justice to the different situation of women. We have already seen that the domestic issue of parenthood influenced the length of time between bachelor's and doctorate degrees more strongly for women than for men. In view of the potentially more varied and formidable obstacles for women scientists, completing the doctorate after a period of part-time study may indicate a particularly high measure of tenacity and determination.

We were slightly puzzled to find that respondents who had worked most closely with their principal dissertation advisor were somewhat more likely to have subsequently left science than those who had not. One might speculate that the respondents who had not worked closely with their dissertation advisor had shown an unusual degree of initiative in finding collaborators better suited to their own interests and thus had formed the beginnings of a collegial network that anchored them in their science careers.

Finally, whereas women who had been research assistants were more likely to quit science than those who had not, the men's propensity for leaving science was less affected by their having been research assistants. The somewhat counterintuitive result that a research assistantship might not be an unequivocally positive factor for women scientists' later career paths may point to unique problems for women in those assistantships. The finding is consistent with Reskin's (1978) hypothesis that some male scientists have difficulty interacting with women as equals and are more comfortable treating them as subordinates. Some male advisors may have tended to allocate the more boring and routine chores to their women research assistants whereas the men assistants received a collegial treatment that would involve them in the more interesting research parts. In such cases, women research assistants would experience research not as an exciting activity but as the exact opposite.

Graduate-study Characteristics and Later Choice of Employment Area

This section only deals with the former postdoctoral fellows who are still active scientists. Respondents who were research assistants were less likely to end up in academe than those who were not (table 3.4). A possible explanation is that research assistantships qualified their holders for nonacademic research positions that are usually higher paying. Moreover, academe might have attracted a much higher proportion of respondents who were teaching assistants and developed an interest in teaching.

Respondents who had been married during graduate school were more likely to work in the nonacademic sector. A less grueling work schedule, more job security, or higher pay might have prompted an overproportionate number of married respondents to become scientists in the nonacademic sector. Respondents with female graduate advisors were also less likely than those with male advisors to work in academe, possibly reflecting the marginal position within academic science of the female advisors as a group.

Whereas the men who had gone straight from their bachelor's degree to their doctorate were more likely to work in academe than those who had interrupted their graduate studies or had worked part time during that phase, the opposite was found for the women. This somehow echoes the gender difference we found in the effect of graduate part-time study on leaving science. Moreover, women with advisors who held a special position of influence tended to go overproportionately into nonacademic science, but this was not true for the men. Association with high-level academic advisors may actually have deterred a number of women from an academic science career. Or, in some cases, high-level advisors might have devoted more energy to furthering their male than their female protégés' academic careers. This is illustrated by a woman interviewee's description of her relationship with her graduate advisor. "Although I had his support, he wasn't tapping into his network for me the way he did for his male grad students. There didn't seem to be, in his mind, the push to have my career be a success." These last two results indicate that the straight, uncomplicated road into academic science was more likely to be traveled by male than by female respondents.

Table 3.4. Graduate-study Predictors of Employment Field

	Full model	Model with significant variables	Interaction model
Controls			
Fellowship	1.37§	1.41§	1.38§
Academic age	0.03*	0.03†	0.03*
Biology	−0.66	−0.72	−0.61
PSME	−0.98†	−0.98†	−0.90†
Time in graduate school			
Time elapsed between degrees	−0.09		
Study time elapsed	0.02		
Part-time study	0.77		
Interruption	0.77		
Part-time study or interruption	−0.83		−1.63‡
Social integration			
Research assistantship	−0.42*	−0.45*	−0.49†
Advisor's rank	0.05		
Advisor's special status	−0.15		0.82
Advisor's gender	−1.12†	−1.05†	−0.97†
Advisor's tenure status	0.41		
Close collaboration with advisor	−0.20		
Collaborative research style	−0.08		
Financing graduate school			
Debt	0.18		
Fellowship	0.22		
Nonuniversity funds (%)	−0.00		
Family status			
Married	−0.43†	−0.34*	−0.32*
Parent	0.27		
Training and performance			
Top university	−0.08		
Publications	−0.16		
Number of publications	0.03		
Gender interactions			
Gender			−0.48
Gender × part-time study or interruption			1.14‡
Gender × advisor's special status			−0.62
N	586	586	614
MCS	103.0	89.4	102.0
Pseudo-R²	0.1379	0.1197	0.1300

NOTES: MCS (model chi-square): reduction in chi-square compared with the chi-square of a baseline model containing only the intercept when the model variables are added. Pseudo-R^2 = [MCS]/[baseline chi-square].
See table 2.2 for explanations of reference marks in this table.

Women may have been less willing to marshal some early advantages toward a straight trajectory into academe; or some of these "advantages" may have been advantageous only for men.

Graduate-study Characteristics and Later Academic Rank

Here we consider only those former postdoctoral fellows who are currently academic scientists. First, we look at academic rank and then at publication productivity.

Respondents with graduate fellowships became more successful in terms of academic rank than those without them (table 3.5). The positive effect of the graduate fellowship may be partly due to the opportunities and the financial support provided by the fellowship, and partly to the selection of talented awardees, partly to a process of self-fulfilling prophecy. In addition, the actual time spent studying toward the doctorate was associated with later academic rank—the shorter the time in graduate school, the higher the current academic rank. Note that it was the actual time spent studying toward the doctorate rather than the years elapsed that predicted academic rank. Obviously, the first variable is a much more precise indicator of ability and motivation than the second. Especially for women, with their higher propensity for interrupting their studies or studying part time, the time elapsed between baccalaureate and doctorate may not be an adequate measure of academic prowess. For men, however, the two time variables coincide more closely. This reinforces our earlier observation of the different effects of absolute time in graduate school on women's and men's subsequent departure from science careers.

For women, a highly collaborative research style during graduate school was not beneficial. On the contrary, it was associated with relatively low academic rank, whereas for men the opposite effect was found. Moreover, women's later rank was somewhat boosted by having had an advisor with a special position, whereas men's academic ranks were not affected by this graduate characteristic. Thus, association with a senior graduate advisor had a positive impact on women's academic careers, but collaboration that was too close had the opposite effect. This finding points to the importance of the type of collaboration. It is consistent with the earlier result that, for our

Table 3.5. Graduate-study Predictors of Academic Rank

	Full model	Model with significant variables	Interaction model
Controls			
Fellowship	0.35‡	0.38§	0.34‡
Academic age	0.08§	0.07§	0.07§
Biology	−0.37†	−0.39‡	−0.37‡
PSME	−0.10	−0.12	−0.06
Affiliation	−0.10	−0.10	−0.07
Time in graduate school			
Time elapsed between degrees	0.02		
Study time elapsed	−0.17†	−0.15§	−0.12§
Part-time study	−0.11		
Interruption	−0.14		
Part-time study or interruption	0.09		
Social integration			
Research assistantship	0.01		
Advisor's rank	0.03		
Advisor's special status	0.12		−0.30
Advisor's gender	0.14		
Advisor's tenure status	−0.22		
Close collaboration with advisor	0.10		
Collaborative research style	0.01		0.25†
Financing graduate school			
Debt	0.16		
Fellowship	0.26†	0.25†	0.24†
Nonuniversity funds (%)	−0.00		
Family status			
Married	0.11		
Parent	0.05		
Training and performance			
Top university	0.14		
Publications	0.07		
Number of publications	0.01		
Gender interactions			
Gender			0.11
Gender × advisor's special status			0.35
Gender × collaborative research style			−0.20‡
N	336	336	348
R^2	0.5184	0.4663	0.5088

NOTE: See table 2.2 for explanations of reference marks in this table.

women respondents, research assistantships correlated with a higher probability of subsequently leaving science. Women who collaborated highly might have been heavily involved in routine tasks and fulfilled subordinate functions in the research process so that their careers did not benefit. Alternatively, even if women did not necessarily play subordinate roles in collaborations, their collaborative research might have been perceived or evaluated as such—which might also have had adverse effects on later rank.

Graduate-study Characteristics and Publication Productivity

The extent of our academic scientists' predoctoral publication productivity was the strongest predictor of their career productivity (table 3.6). For instance, the respondents with no predoctoral publications produced an annual average of 1.9 publications as compared with 3.0 publications among those with predoctoral publications. It was the men in particular who followed the simple pattern "the more productive in graduate school, the more productive overall"; the women's overall publication record did not seem to benefit much from a small number of predoctoral publications. Again, the fact that women's productivity record was somewhat less predictable from its beginnings may indicate that the their publication history was more subject to all sorts of unexpected influences so that there was a less steady accumulation of initial advantages.

The actual time of graduate study also had its expected effect. The faster the respondents had gone through graduate school, the more productive they became. Surprisingly, women who had held graduate fellowships published less than those without them. On the other hand, a fellowship had, if anything, a slight positive effect on the men's average career productivity. One might speculate that women, to a higher degree than men, considered the fellowship as a reward in itself that would partly substitute for publication productivity in conveying career advantages.

Marriage and parenthood had opposite impacts: those who were married during graduate school turned out to become more productive than those who were not, whereas graduate students with children had a lower career productivity than those without. Apparently, a marriage during graduate school may have conveyed

Table 3.6. Graduate-study Predictors of Publication Productivity

	Full model	Model with significant variables	Interaction model
Controls			
Fellowship	0.09	0.09	0.11
Academic age	0.00	0.01	0.01
Biology	0.10	0.13	0.13
PSME	0.10	0.13	0.13
Time in graduate school			
Time elapsed between degrees	−0.04		
Study time elapsed	−0.04	−0.07‡	−0.08§
Part-time study	0.06		
Interruption	0.40*		
Part-time study or interruption	−0.35		
Social integration			
Research assistantship	0.01		
Advisor's rank	−0.04		
Advisor's special status	−0.01		
Advisor's gender	0.16		
Advisor's tenure	0.13		
Close collaboration with advisor	−0.03		0.40*
Collaborative research style	0.04		
Financing graduate school			
Debt	−0.00		
Fellowship	0.03		0.45†
Nonuniversity funds (%)	0.00		
Family status			
Married	0.19‡	0.17‡	0.17†
Parent	−0.18†	−0.20†	−0.20†
Training and performance			
Top university	−0.07		
Publications	0.15*	0.16†	0.64‡
Number of publications	0.06§	0.06§	0.06§
Gender interactions			
Gender			0.72‡
Gender × close collaboration with advisor			−0.32†
Gender × fellowship			−0.35†
Gender × publications			−0.35†
N	324	324	348
R^2	0.2998	0.2744	0.5088

NOTE: Dependent variable is a natural logarithmic transformation of annual publication productivity in order to increase the normality of this variable. See table 2.2 for explanations of reference marks in this table.

the advantages of a social support system, but children may have diluted energy and concentration. Both findings about the long-term career repercussions of marriage and parenthood during graduate school indicate that the timing of marriage and of starting a family is important—not only for women but also for men.

Interestingly, women who had collaborated most closely with their actual supervisor became less productive than those who had not. In contrast, this graduate-study characteristic had, if anything, a positive impact on men's career publication record. A distant principal advisor may have enabled women to become productive on their own by strengthening their sense of independence and resourcefulness in conducting their research.

Together, the various findings about different aspects of collaboration during graduate school highlight the complex potential problems and pitfalls lurking in this area for women. By contrast, collaboration was found much more unproblematic for the men respondents. On the whole, their careers appeared to be robust—that is, proceeded in the same average trajectories—regardless of the particulars of their graduate experience with advisors and collaboration. In terms of rank, it was beneficial for women graduate students to have a high-ranking advisor but not to get lured into heavy collaboration, which seemed to carry disproportionately little benefit for women. Moreover, women who were distant from their advisor became more productive than those who were close. At the graduate level, the problem appeared to lie less in women's isolation from senior scientists than in the nature of women's collaboration with them.

Postdoctoral Fellowship

After a scientist's formal training has come to its conclusion with the award of the doctorate, the postdoctoral fellowship is a crucial phase during which important decisions for the further career take place. This section examines the career stage of the postdoctoral fellowship and its repercussions on later career outcomes. It particularly focuses on gender differences in the postdoctoral experience and investigates whether any such postdoctoral gender differences constitute early gender advantages or disadvantages because they translate into differences in later career outcomes. Because the award of a prestigious postdoctoral fellowship was the selection criterion for our sample, the respondents' postdoctoral experience was the career stage in which their career paths started fanning out into different directions.

A large-scale National Research Council study of postdoctoral fellowships (NRC 1981, 227) concluded that it "found no difference in the way women and men approach postdoctoral education." The NRC study, however, addressed the gender issue only in passing. Our more detailed approach to potential postdoctoral gender differences will endeavor to discover such differences that reveal themselves on a closer look—even in our sample, in which there may be fewer gender differences than in the general population of postdoctoral fellows.

We selected a number of postdoctoral characteristics that appeared likely to have an impact on later career outcomes. In the first step, we examined these characteristics mainly for gender differences. In the second step, we examined whether the postdoctoral characteristics were statistically related to career outcomes. We also investigated whether any postdoctoral characteristics affected career outcomes in different ways for the two genders.

Function of Fellowship

The first group of postdoctoral variables concerned the function of the postdoctoral fellowship. The principal difference here lay between a qualifying and a holding function (Reskin 1976). On the one hand, the postdoctoral fellowship represents a transitory career stage that rewards good performance in graduate school and qualifies a young scientist for an academic career at prestigious institutions. On the other hand, scientists who cannot get the kind of position they want after receiving their doctorate may use a postdoctoral fellowship as a holding position that allows them, at least temporarily, to maintain a marginal foothold in research science. Respondents were deemed to have taken their fellowship for its qualifying function if they said they took it to obtain additional research experience, work with a particular scientist or research group, or finish a book or other major project.

To probe for the holding function of the postdoctoral fellowship, the questionnaire asked the respondents whether they took their NSF or NRC postdoctoral fellowship because they could not obtain the position they wanted. In addition, we asked whether the respondent prolonged any postdoctoral fellowship and, if yes, whether the fellowship was prolonged because of difficulties in finding other employment.

Another common function of a postdoctoral fellowship is to help broaden one's research field or to switch fields. This may have a slightly negative, if any, effect because it constitutes a deviation from a straight career path. We asked whether our respondents took the fellowship to switch into a different field of research and whether they broadened or changed fields while being a fellow. Finally, we inquired whether the respondents took their fellowship to be with their spouses or other close persons.

Duration of Fellowship

The second group of postdoctoral variables was connected with the duration of the fellowship phase: number of postdoctoral applications, number of successful applications, and total months spent in postdoctoral appointments. We also noted whether the NSF or NRC postdoctoral fellowship immediately followed graduation (that is, began in the same calendar year in which the doctorate was received) or whether it started at least one year later.

Social Integration

The third group of postdoctoral variables described science-internal characteristics that related to the fellows' integration into their institutional environment. Such variables included the number of postdoctoral mentors, the senior or junior status of the postdoctoral advisor, and the advisor's gender. The respondents were also asked whether there were other scientists, apart from their advisor, with whom they worked closely or who otherwise significantly influenced their work during the fellowship. Furthermore, our respondents reported their degree of scientific collaboration during the postdoctoral fellowship on a five-point scale—ranging from 1 ("usually worked alone on my scientific problems") to 5 ("my research was performed almost exclusively in collaboration with other scientists"). Finally, we wondered whether the respondents were able to devote to their fellowship the level of attention they thought was needed. Our measure distinguished those who either did not think full attention was necessary or who could give full attention from those who could not give full attention although they thought it was necessary. In other words, this variable differentiated those who thought their level of involvement in their postdoctoral fellowship was satisfactory from those who perceived a problem in this area.

Training and Performance

In this fourth category, we recorded whether the respondents spent their fellowship at a university or nonuniversity institution and whether they attended a university department that was rated among the top 15 percent of research-doctorate departments (according to Jones, Lindzey, and Coggeshall 1982). We further noted whether the respondents had any postdoctoral publications at all and calculated the monthly postdoctoral-publication rate as a measure of postdoctoral-publication productivity.

Family Status

The final group of characteristics pertained to the science-external family sphere. Here the variables were marital status and parental status as well as becoming a parent during the fellowship.

95

Table 3.7. Means of Postdoctoral Characteristics by Gender

Variable	Men	Women
Function of postdocorate		
Qualifying function	0.942	0.840§
No other job available	0.123	0.138
Prolonged postdoctorate	0.406	0.416
Prolonged because of lack of job	0.130	0.136
Took postdoctorate for field switch	0.229	0.287
Changed research field during postdoctorate	0.633	0.586
Took postdoctorate to be with spouse	0.085	0.245§
Duration of postdoctorate		
Number of applications	1.80	1.82
Number of successful applications	1.65	1.67
Months as fellow	29.50	30.66
Time lag between Ph.D. and postdoctorate	0.444	0.421
Social integration		
Number of mentors	1.53	1.53
Junior advisor	0.118	0.164
Senior advisor	0.761	0.720
Male advisor	0.855	0.831
Female advisor	0.032	0.063
Outside collaborator	0.680	0.640
Collaborative research style	2.55	2.32†
No full attention to postdoctorate	0.067	0.127†
Training and performance		
University	0.613	0.441§
Top university	0.359	0.234‡
Publications	0.817	0.730†
Publication rate	0.23	0.17‡
Family status		
Married	0.661	0.637
Parent	0.412	0.302‡
Child born during postdoctorate	0.137	0.090*

NOTE: See table 2.2 for explanations of reference marks in this table.

Summary of Postdoctoral Gender Differences in Means

Table 3.7 displays the mean of each of the twenty-four postdoctoral variables by gender. It gives a rough idea of gender differences in the postdoctoral experience.

We selected awardees of the NSF and NRC postdoctoral fellowships as our population in order to maximize the proportion of postdoctoral fellows who took the fellowship for its qualifying function. Indeed, a large majority of our respondents in all fields took the fellowship for this purpose. Only a small group noted the absence of other employment possibilities as a reason for prolonging their fellowships, and there were no gender differences (men: 12.3 percent; women: 13.8 percent). In the general population of postdoctoral fellows, this group is about twice as large (28.6 percent; NRC 1981, 135)—another indication of the elite status of the NRC and NSF fellowship compared with postdoctoral fellowships in general.

Overall, the picture that emerged from the various postdoctoral characteristics was one of gender similarity rather than difference. Significant gender differences were notably absent in the areas of fellowship duration and the advisor's rank and gender. It may well be that recipients of NSF and NRC postdoctoral fellowships were intrinsically appealing to potential mentors and that any disadvantage in this respect that may exist for other female scientists (who did not receive an NSF or NRC postdoctoral fellowship) did not apply to our select group of promising young scientists.

Nevertheless, in contrast with the cited NRC finding of no gender differences in the postdoctoral experience (NRC 1981, 227), our study did discover significant differences between the genders in the following respects:

Women took the postdoctoral fellowship more often to be with their spouse and less frequently for career-enhancing reasons.
Fewer women than men were parents while doing their postdoctoral fellowship.
Fewer women than men could give full attention their postdoctoral research, although they thought it was necessary.
Women, on the average, had a weaker record of publications from their fellowship than men did.

Women's postdoctoral research style was less collaborative than men's.

Women also seemed less likely than men to have been postdoctoral fellows at universities and at top graduate departments. This gender difference disappeared after controlling for type of fellowship, academic age, and academic fields.

The gender similarity in marital status among our respondents may not reflect the situation among doctoral scientists in general. Among all American doctoral scientists, those who entered a postdoctoral fellowship were found to be less likely to be married than those who entered other job positions. This was particularly true for men (NRC 1981, 149).

Outcomes

We will now examine whether these postdoctoral characteristics had any statistically measurable impact on key career outcomes (those already discussed in the graduate-school section).

Postdoctoral Characteristics and Later Leaving Science

The longer the postdoctoral phase lasted, the higher the tendency to persist in science (table 3.8). More precisely, a long duration of the postdoctoral fellowship, as indicated by a larger number of postdoctoral applications, was associated with persistence in science for women to a higher extent than for men. Apparently, women in science have a somewhat higher propensity to remain in marginal postdoctoral positions. This may indicate their greater tenacity in keeping a foothold in science, or perhaps a greater opportunity to do so, because their husbands were typically the main breadwinners of the family.

Fellows who later chose careers unrelated to research science were also more likely to have become parents during the fellowship. This may be a small indication of problems combining family and career. The postdoctoral fellowship appeared to have been a particularly inopportune time for becoming a parent in terms of a later career in science.

Table 3.8. Postdoctoral Predictors of Leaving Science

	Full model	Model with significant variables	Interaction model
Controls			
Fellowship	0.20	0.09	0.14
Academic age	0.03	0.04*	0.04†
Biology	−0.09	−0.10	−0.26
PSME	−0.63	−0.69	−0.68
Function of postdoctorate			
Qualifying function	1.00		
No other job available	0.10		
Prolonged postdoctorate	−0.23		
Prolonged because of lack of job	0.63		
Took postdoctorate for field switch	0.28		
Changed research field during postdoctorate	−0.12		
Took postdoctorate to be with spouse	−0.52		
Duration of postdoctorate			
Number of applications	−0.33		1.65†
Number of successful applications	0.35		
Months as fellow	−0.04†	−0.03‡	−0.02*
Time lag between Ph.D. and postdoctorate	0.23		
Social integration			
Number of mentors	0.08		
Junior advisor	0.73*	0.65*	0.69*
Male advisor	−0.09		
Female advisor	−0.07		55.36§
Outside collaborator	−0.33		
Collaborative research style	0.09		
No full attention to postdoctorate	0.45		
Training and performance			
University	0.11		
Top university	−0.31		
Publications	0.02		
Publication rate	−0.74		
Family status			
Married	0.19		
Parent	0.41		
Child born during postdoctorate	0.54	0.74*	0.47

Table 3.8. *Continued*

Gender interactions			
Gender			2.10[†]
Gender × number of applications			−1.61[†]
Gender × female advisor			28.35[†]
N	614	614	675
MCS	31.9	20.6	35.98
Pseudo-R²	0.1014	0.0595	0.0899

NOTES: MCS (model chi-square): reduction in chi-square compared with the chi-square of a baseline model containing only the intercept when the model variables are added. Pseudo-R^2 = [MCS]/[baseline chi-square].
See table 2.2 for explanations of reference marks in this table.

Fellows affiliated with junior scientists had a higher rate of leaving science than those affiliated with senior scientists. This is what one would expect because a senior advisor may be able to smooth the postdoctoral fellow's entry into the science system to a higher degree than a junior advisor could. The effect of the advisor's gender, however, was very surprising. Women who were affiliated with female advisors had a *higher* rate of later leaving science than those who were not (16.7 percent versus 9.7 percent). For men, the reverse was the case. None of those affiliated with a female advisor, as compared with 8.7 percent of the others, left science. This casts doubt on a simple (positive) mentor or role-model effect of women advisors on women scientists' careers and will be discussed later in greater detail.

Postdoctoral Characteristics and Later Choice of Employment Area

The following section focuses on the former postdoctoral fellows who are still active scientists. The postdoctoral holding function was associated with subsequent nonacademic employment (table 3.9). Problems in finding an easy entry into a professorial career, as indicated by employment difficulties after the fellowship, may have finally translated into the acceptance of nonacademic positions. The respondents who had an interval between their Ph.D. graduation and the start of their fellowship were somewhat more likely to stay in academic science than those who went to the fellowship immediately after graduation. Taking a fellowship after a pause may be less automatic than going straight to the fellowship after the completion of the doctorate. Apparently, those who decided to take the fellowship after an interval based their decision on a particularly strong commitment to academic science.

Initially, a somewhat puzzling finding was that both variables describing postdoctoral publication activity predicted the tendency of later choice of employment field—however, in opposite ways. On the one hand, a high postdoctoral publication rate predicted a later position in academe. On the other hand, the former fellows who had no postdoctoral publications at all were also more likely to be currently in academe than those who had any such publications. In other words, both the former fellows with a prolific postdoctoral publication record and the former fellows with no postdoctoral publications at all gravitated toward academe, whereas those with a moderate number of publications tended to become nonacademic scientists.

It seemed obvious why those with a strong postdoctoral publication record tended to have academic jobs: as we have seen, publication productivity is the major performance indicator for an academic scientist. It was less obvious, however, why those with no postdoctoral publications at all were also drawn to academe. A plausible explanation would be that academic jobs offer a much wider variety in research intensity than nonacademic science jobs. While most positions at prestigious universities place an overriding emphasis on research, positions at undergraduate institutions typically entail teaching. Because nonacademic science jobs usually lack the teaching component, they tend to be more strongly oriented toward research than the latter kind of academic jobs. Thus, former fellows with an inclination to teach may have gravitated toward academic teaching positions. And for scientists who wanted to avoid research, academic teaching jobs may have provided a more effective shelter from research than nonacademic science jobs would have been able to do.

Women with nonadvisorial collaborators were more likely to work in academe than those without a nonadvisorial collaborator, but we did not find such a pattern among the men. This gender

Table 3.9. Postdoctoral Predictors of Employment Field

	Full model	Model with significant variables	Interaction model
Controls			
Fellowship	1.05‡	1.44§	1.41§
Academic age	0.02	0.03*	0.02
Biology	−0.44	−0.46	−0.53
PSME	−0.83*	−0.81*	−0.85†
Function of postdoctorate			
Qualifying function	0.12		
No other job available	−0.20		
Prolonged postdoctorate	−0.15		
Prolonged because of lack of job	−0.85†	−0.88‡	−0.95§
Took postdoctorate for field switch	−0.35		
Changed research field during postdoctorate	0.17		
Took postdoctorate to be with spouse	−0.22		
Duration of postdoctorate			
Number of applications	0.28		
Number of successful applications	−0.26		
Months as fellow	−0.00		
Time lag between Ph.D. and postdoctorate	0.23	0.35*	0.36*
Social integration			
Number of mentors	0.16		
Senior advisor	−0.20		
Male advisor	0.12		
Female advisor	0.15		
Outside collaborator	−0.08		−0.45
Collaborative research style	0.08		
No full attention to postdoctorate	−0.38		
Training and performance			
University	0.39		
Top university	0.17		
Publications	−0.72†	−0.67†	−0.75‡
Publication rate	1.13†	1.29†	1.32‡
Family status			
Married	−0.24		
Parent	0.32		
Child born during postdoctorate	−0.15		

Table 3.9. *Continued*

Gender interactions			
Gender			−0.50
Gender × outside collaborator			0.44
N	564	564	613
MCS	115.1	99.8	107.8
Pseudo-R²	0.1580	0.1370	0.1375

NOTES: MCS (model chi-square): Reduction in chi-square compared with the chi-square of a baseline model containing only the intercept when the model variables are added. Pseudo-R² = [MCS]/[baseline chi-square].
See table 2.2 for explanations of reference marks in this table.

difference accentuates the importance of establishing a collegial network during the postdoctoral fellowship for women's persistence in academic science.

Postdoctoral Characteristics and Later Academic Rank

In the following we deal with only those former postdoctoral fellows who are currently academic scientists.

Spending a longer period of time in postdoctoral positions lowered average rank (table 3.10). This duration effect was also reflected in a related characteristic—the number of postdoctoral applications. Controlling for duration, however, we noted that the number of *successful* applications correlated with high academic rank. To some extent, being awarded with postdoctoral fellowships probably indicated that the applicants possessed qualities that made them well suited for an academic career.

A strong postdoctoral publication record presaged higher rank achievement, as one might expect. Furthermore, scientists who took the fellowship because other positions were unavailable held an lower average rank than those who did not give that reason. Thus, using the fellowship for its holding function adversely affected subsequent academic rank. Taking the fellowship for its qualifying function, by contrast, had the opposite effect—but only for men. In this instance, men as a group were better able to reap the benefits of the fellowship's qualifying function.

An important finding is that taking their fellowship to be with their spouse put scientists at a disadvantage in terms of academic rank. This effect applied similarly to women and men in our sample. The vast majority of the former fellows who gave this motivation for their fellowship also indicated that they took it for its qualifying function. That is, they typically presented their decision to take the postdoctoral fellowship as a combination of intrinsic-scientific and external-familial considerations. This juggling act between scientific and familial motivations resulted, as noted, in subsequently lower average academic rank. In some cases, giving an additional science-internal motivation for taking the fellowship may have been a benign camouflage of the overriding importance of familial considerations. Of course, the fact that a prestigious postdoctoral fellowship is hardly devoid of any scientific value facilitates the additional mention of science-internal motivations. Nonetheless, taking the postdoctoral fellowship to be with one's spouse had negative career repercussions—be it because the location of the fellowship in itself was of crucial importance for the fellow's later academic career or because following one's spouse for the fellowship was part of a more general pattern reflecting a couple's fundamental decision about whose career has priority.

Becoming a parent during the postdoctoral fellowship was found to have a negative impact on later rank for the women. The men's rank, by contrast, was less affected. Thus, the postdoctoral fellowship may be an unfortunate time for women academics to have babies.

An initially surprising finding was that fellows who were associated with an advisor of junior rank later came to hold a somewhat higher average rank than those who did not. Thus, contrary to what we had expected, an association with a senior postdoctoral advisor may not have been unequivocally advantageous. Whereas senior male advisors may be able to further their advisees' careers through their old boys' network of contacts, they may not always be inclined to do so, and they may also be able to use their postdoctoral fellows' labor in ways that do not benefit the fellows' later careers. On the other hand, our finding appears to support the suggestion that it is advantageous for scientists to have mentors between ten and twenty years their senior (Simonton 1988, 114–115). Affiliation with junior

Table 3.10. Postdoctoral Predictors of Academic Rank

	Full model	Model with significant variables	Interaction model
Controls			
Fellowship	0.51‡	0.42§	0.43§
Academic age	0.08§	0.08§	0.08§
Biology	−0.24•	−0.26•	−0.28†
PSME	−0.04	−0.06	−0.04
Affiliation	−0.07	−0.07	−0.07
Function of postdoctorate			
Qualifying function	0.07		1.02†
No other job available	−0.27•	−0.28†	−0.31†
Prolonged postdoctorate	−0.04		
Prolonged because of lack of job	−0.03		
Took postdoctorate for field switch	−0.02		
Changed research field during postdoctorate	0.04		
Took postdoctorate to be with spouse	−0.23•	−0.24†	−0.18
Duration of postdoctorate			
Number of applications	−0.12	−0.13•	−0.14•
Number of successful applications	0.24†	0.26‡	0.25‡
Months as fellow	−0.02§	−0.02§	−0.01§
Time lag between Ph.D. and postdoctorate	0.08		
Social integration			
Number of mentors	−0.03		
Junior advisor	0.23	0.26†	0.25†
Senior advisor	−0.04		
Male advisor	0.03		
Female advisor	0.10		
Outside collaborator	0.07		
Collaborative research style	0.01		
No full attention to postdoctorate	−0.21		
Training and performance			
University	−0.14		
Top university	0.04		
Publications	0.01		
Publication rate	0.41‡	0.43§	0.45§

Table 3.10. *Continued*

Family status			
Married	0.09		
Parent	−0.02		
Child born during postdoctorate	−0.23	−0.21˙	0.35
Gender interactions			
Gender			0.42
Gender × qualifying function			−0.62˙
Gender × child born during postdoctorate			−0.46
N	310	310	346
R^2	0.5834	0.5748	0.5760

NOTE: See table 2.2 for explanations of reference marks in this table.

advisors may be a strategic advantage, especially if they are rising scientific stars. Such association would acquaint the postdoctoral fellows with future scientific leaders and at an early time familiarize them with new ideas and methods that are going to become influential in their discipline. If we consider this finding together with the previously noted result that postdoctoral fellows with junior advisors were more likely subsequently to leave science, an association with junior advisors appeared to polarize outcomes for postdoctoral fellows. They either tended to leave science altogether or to become very successful academic scientists. Affiliation with a junior advisor may thus be considered as something of a high-risk move—with increased opportunities but also increased pitfalls.

As we have already noted, women were slightly less collaborative in their research style than men during their postdoctoral fellowship. But higher postdoctoral collaboration appeared to predict a worse career outcome in terms of rank for women and a better outcome for men, although this gender interaction was not quite statistically significant ($p = 0.1153$). This hints at differences in the nature of collaboration for women and men postdoctoral fellows. At the postdoctoral level, as at the graduate level, the problem seems to be less related to women's isolation from collaboration altogether and more related to the quality and consequently the benefits of col-

Table 3.11. Rank by Gender, Degree of Postdoctoral Collaboration, and Type of Collaborator

	Women			Men		
	All	*No NAC*	*NAC*	*All*	*No NAC*	*NAC*
Low collaboration	2.7	2.6	2.7	2.9	2.9	3.0
N	122	45	77	252	105	147
High collaboration	2.4	2.0	2.6	3.1	3.1	3.1
N	63	22	41	232	52	180

NOTES: NAC = nonadvisorial collaborator. If the residual of a regression of collaboration style on type of fellowship, year of fellowship, and fields during fellowship was 0 or positive, the respondent joined the high-collaboration group. Those with negative residuals formed the low-collaboration group. Ns include all fellows.

laboration. The issue is further complicated by the type of postdoctoral collaborators.

As we mentioned, almost as many women (64.0 percent) as men (68.0 percent) postdoctoral fellows reported that they worked closely with or were influenced by scientists other than their advisors. Among the men, those who collaborated with scientists other than their mentors reported a more collaborative overall research style than those who had no such contacts ($r = 0.23$). Put differently, only 58.3 percent of the male low collaborators, but 77.6 percent of the male high collaborators, worked with scientists other than their formal postdoctoral advisors. For women, however, this relationship hardly existed. Among them, similar percentages of the low collaborators (63.1 percent) and the high collaborators (65.1 percent) worked with scientists other than their advisors. Thus, in the group of highly collaborative fellows, disproportionately many women collaborated only with their mentors, while men were more likely to work also with other scientists.

For highly collaborative women the absence of nonadvisorial collaborators appeared particularly detrimental. They held an average rank of only 2.0, as compared with average ranks of 2.6 for both highly collaborative women with nonadvisorial contacts and less collaborative women without nonadvisorial collaborators (table 3.11). This underscores both the potentially problematic nature of women's close collaboration with their advisor and the importance of forming collegial networks through contacts with scientists other than their advisor.

Postdoctoral Characteristics and Publication Productivity

Not surprisingly, postdoctoral publication activity predicted career publication productivity the most strongly (table 3.12). For instance, the respondents with no postdoctoral publications produced an annual average of 1.7 publications compared with 3.0 publications produced by former fellows who did have postdoctoral publications. Overall publication productivity, however, was more sensitive to the postdoctoral productivity rate among men than among women. The women's relatively weak coupling of postdoctoral and career productivity echoed the earlier gender difference in the relationship between predoctoral publications and career prodictivity. Male respondents' overall productivity also benefited from spending the postdoctoral fellowship at a highly prestigious department, whereas the opposite appeared to be true for the female respondents.

Former fellows who did not enter their fellowships immediately after receiving the doctorate became less productive on the average than those who went straight from graduate school to the postdoctoral fellowship—another indication of the advantages of a straight career path. Respondents who were married during the fellowship had a slightly higher overall productivity than those were not. Here, the social support and stability provided by marriage may have contributed to the positive long-term effect on productivity.

Men postdoctoral fellows who were affiliated with senior advisors became more productive than men who were not (3.0 versus 2.5). But we were startled to find the reverse was true for women fellows (2.1 versus 2.7). Women's productivity thus appeared to be adversely affected by association with a senior advisor. The somewhat counterintuitive result might indicate that the old boys' network indeed works best for "boys." This finding supports an emerging theme: women postdoctoral fellows were not necessarily isolated from influential members of their profession. Rather, on average they derived fewer benefits from such collaboration than men did, which points to a different quality of interaction. Possibly, the women were not as often given research work that would be useful in their own

Table 3.12. Postdoctoral Predictors of Publication Productivity

	Full model	Model with significant variables	Interaction model
Controls			
Fellowship	−0.03	0.07	0.04
Academic age	−0.00	0.00	0.00
Biology	0.22*	0.24†	0.24†
PSME	0.21*	0.26†	0.22†
Function of postdoctorate			
Qualifying function	0.01		
No other job available	0.10		
Prolonged postdoctorate	0.12		
Prolonged because of lack of job	−0.21		
Took postdoctorate for field switch	−0.04		
Changed research field during postdoctorate	0.07		
Took postdoctorate to be with spouse	−0.14		
Duration			
Number of applications	−0.06		
Number of successful applications	0.11		
Months as fellow	−0.00		
Time lag between Ph.D and postdoctorate	−0.14*	−0.16†	−0.17‡
Social integration			
Number of mentors	0.02		−0.15
Junior advisor	1.08†		
Senior advisor	1.02†		0.72‡
Male advisor	−0.97†		
Female advisor	−0.72		
Outside collaborator	−0.03		
Collaborative research style	0.01		
No full attention to postdoctorate	−0.13		
Training and performance			
University	0.13		
Top university	0.03		0.36*
Publications	0.31‡	0.31§	0.29§
Publication rate	0.55§	0.59§	1.05§
Family status			
Married	0.16*	0.13*	0.15†

Table 3.12. *Continued*

Parent	−0.12		
Child born during postdoctorate	0.15		
Gender interactions			
Gender			0.27
Gender × senior advisor			−0.56‡
Gender × number of mentors			0.14*
Gender × top university			−0.26*
Gender × publication rate			−0.33
N	308	308	322
R²	0.2613	0.2089	0.2740

NOTE: Dependent variable is a logarithmic transformation of annual publication productivity.
See table 2.2 for explanations of reference marks in this table.

later careers, or some of the advisors tried harder to advance their men than their women protégés' careers. From a policy point of view, it may not be sufficient to advocate that women postdoctoral fellows be assigned to senior members of the department. The focus of attention should be on improving or monitoring the quality of advisorship that women receive. In a related result, a higher number of postdoctoral mentors correlated positively with career productivity for women but negatively for men. This underscores the potential dangers of women's association with a single mentor.

Gender and the Accumulation of Human Capital

We now discuss the effects of graduate and postdoctoral characteristics in conjunction, placing them within our theoretical framework. We found significant relations between graduate and postdoctoral characteristics and later career outcomes. This supports the notion that science careers are sensitive to small initial differences and the accumulation of advantages and disadvantages. The size of the observed effects was generally small, however, probably due in part to the fact that we were dealing with a severely restricted range of scientists—those who had all received a prestigious postdoctoral fellowship. In our sample, the shared characteristic of the award of a prestigious fellowship may have outweighed individual differences in graduate school and postdoctoral fellowship experiences. As a rough measure of the amplitude of the effects of graduate and postdoctoral characteristics on career outcomes, we used the increase in the explained variance that resulted from entering the significant graduate or postdoctoral characteristics into a regression model that previously only contained the controls. Adding the significant graduate or postdoctoral variables to such a model explained not more than 5 percent extra variance in most cases.

A noteworthy exception was found in the area of publication productivity. There, graduate characteristics explained an extra 25.9 percent, and postdoctoral characteristics explained an extra 19.4 percent of the variance in career publication productivity—mainly thanks to the predictive power of the predoctoral (and postdoctoral fellowship) publication records. Clearly, respondents who got into the habit of publishing early tended to accumulate a prolific career publication record. The area of publication productivity thus emerged as the one most sensitive to the accumulation of advantages and the amplification of small initial differences. Whereas other outcomes, such as academic rank, appear to be subject to a variety of different tendencies that in some instances might also cancel each other out, publication productivity simply seems to beget more publication productivity.

The accumulation of graduate and postdoctoral advantages and disadvantages differed by gender in important ways. We now briefly recapitulate our results within the typology of gender differences and disadvantages that we have already outlined. We note again that we are dealing with statistical tendencies, not with strict separation lines between the genders.

The first type of postdoctoral effect that we distinguished is one of gender-neutral advantages and disadvantages (type 1). For the graduate phase these included being married during graduate school; having been awarded a graduate fellowship; having a predoctoral publication record; working most closely with the principal dissertation advisor; and the actual time spent in graduate studies. In the postdoctoral phase, we found the following type 1 effects: the duration of the fellowship (with the exception that a long duration predicted remaining in science for women in particular); taking the fellowship in a holding pattern; having a gap between receiving the doctorate and entering the postdoctoral fellowship; affiliation with advisors of junior status; and being married as a fellow. All these characteristics were shared by similar proportions of male and female respondents and had similar repercussions for their later career paths.

By contrast, type 2 advantages and disadvantages were graduate and postdoctoral characteristics that were shared by the genders in similar proportions but that had different repercussions for women's and men's careers. In other words, the groups of women and men in these cases had the same amount of human capital but obtained different interest rates—women typically were less able to capitalize on these characteristics. In this category, we found having a collaborative research style during graduate school; having an advisor of influential status; working most closely with the principal advisor; and having a predoctoral publication record. Among the postdoctoral characteristics were the location of the postdoctoral fellowship; affiliation with a female advisor and a senior advisor; and collaborating closely with a scientist other than one's advisor. The number of postdoctoral mentors also belongs in this category.

Type 3 advantages and disadvantages were graduate and postdoctoral gender differences that translated into gender disadvantages. For instance, more women interrupted their graduate studies, and such interruption was associated with later leaving science. Women had a smaller average number of predoctoral publications, which was associated with lower career publication productivity. In addition, more women than men took the postdoctoral fellowship to be with their spouse, which had negative effects on later academic rank. In these cases, a gender difference at an earlier stage was transformed into a gender disadvantage at the outcome level.

Conversely, fewer women graduate students had children; and having been a parent in graduate school was negatively associated with career publication productivity. Similarly, becoming a parent during the postdoctoral fellowship was associated with a higher propensity for leaving science; and somewhat fewer women than men became parents during the fellowship. In these instances, women as a group appeared to have avoided a characteristic with a deleterious career effect. One may speculate that the higher proportion of women respondents without children during graduate school and postdoctoral fellowship reflected women's anticipation of the difficulties of combining a science career and a family. Many women may have realized that having a family during the graduate school or postdoctoral phases could cause negative repercussions for their scientific careers, so that delaying (or forgoing) parenthood appeared to be the rational choice. This indicates that our women respondents should not be seen as victims of social trends beyond their control. There was room for choices, and our women respondents made some prudent ones—as one might expect from our sample of people who were particularly successful in graduate school and thus received a prestigious postdoctoral fellowship.

Type 4 describes graduate and postdoctoral gender differences with gender-specific outcome effects. Predoctoral as well as postdoctoral publication productivity were key characteristics on which women were at a double disadvantage. Women, for instance, had fewer postdoctoral publications than men did; and a strong postdoctoral publication record had a stronger positive effect on men's than women's career publication record. Similarly, fewer women than men respondents said they took the postdoctoral fellowship for its qualifying function, and this motivation had a stronger positive impact on men's than on women's later careers. In contrast with this compounding of gender disadvantages (less capital *and* lower interest rates), a number of other results illustrated the compensation concept. The women in our sample—successful survivors of the graduate phase and awardees of a prestigious postdoctoral fellowship—did a relatively good job in accumulating human capital and steering clear of debts. As a group, they seemed to have correctly anticipated difficulties from certain graduate and postdoctoral characteristics and consequently avoided them—or to have made the best out of a less than ideal starting position. For instance, fewer women than men were graduate research assistants; and for women, being a research assistant predicted a higher rate of leaving science. Similarly, slightly fewer women than men became parents during the postdoctoral fellowship—a postdoctoral characteristic that predicted a lower average rank, particularly for women. Moreover, it took women as a group longer than men to receive their doctorate; but a long period between bachelor's and doctorate degrees indicated a greater persistence in science for women, whereas for men it indicated a greater propensity to leave science.

Finally, we should add a technical note of caution, especially regarding types 3 and 4. Correlational techniques of the kind used do not imply causality and are liable to show spurious effects. If, for example, women and men differ in a certain graduate-study characteristic and a regression shows that this characteristic is connected with later career outcomes, the characteristic is not necessarily a gender disadvantage—that is, a gender difference that negatively affects later career outcomes. Reversely, one could argue that because women's career outcomes are worse than men's for some other reason (possibly discrimination at a later stage), any characteristic associated with women may *appear* to affect career outcomes negatively, even though it does not. In other words, by including more and more human capital variables, one could eventually make the gender main effect statistically disappear.

Two Themes

Gender differences in terms of the possession and accumulation of human capital were concentrated in two areas in which highly complex—and in some respects surprising—pictures of gender differences and disadvantages emerged. These areas were the impact of domestic circumstances and collaboration and interaction with advisors.

The Interpenetration of Family and Career

We found stronger links between family status and science career for our women than for our men respondents. For instance, women's family responsibilities radiated into some of their postdoctoral characteristics, whereas men's family status hardly affected those characteristics. For women, but not for men, the inability to devote one's whole attention to research was evidently correlated with domestic circumstances during the postdoctoral fellowship. Among the women, 6.0 percent of the unmarried postdoctoral fellows but 16.5 percent of the married fellows could not devote their full energy to their research, although they thought it was needed. For men, marital status did not make any difference. Of the female fellows without children, 8.4 percent could not devote their full energy to their research, although they thought it was needed. By contrast, a proportion almost three times as large (23.2 percent) of the female fellows with children gave this response. For men, being a parent did not make any difference in this respect. Thus, whereas family responsibilities appeared to handicap the women postdoctoral fellows' research effort, they did not at all affect the men's, on the average.

As we pointed out earlier, women respondents took their fellowship more often to be with their spouse than did men; but there was also a gender difference in the effect of children on the motivation of taking the fellowship for this reason. Women were more likely to take the fellowship because of their spouse if they had children. Of the women who had children at the time of their postdoctoral fellowship, 30.1 percent took the fellowship to be with their spouse or another close person, compared with 22.0 percent of the women without children. For the men the opposite effect was observed. Only 3.0 percent of men with children, as compared with 12.2 percent of men without children, named being with their spouse as a reason for taking the fellowship. The presence of children appeared to assign priority to the husband's career over the wife's career, and the pursuit of the wife's professional opportunities may have lost importance relative to child-raising responsibilities. Hence, the presence of children crystallized the traditional division of labor between husband and wife.

A few domestic obligations did have the expected negative impact on career outcomes. Postdoctoral fellows who became parents during the fellowship were more likely to leave science. Respondents who took the fellowship to be with their spouses achieved lower average ranks, as did women who had babies during the fellowship. On the other hand, being married during the postdoctoral fellowship had a *positive* effect on career publication productivity. The effects of marriage were thus not unequivocal. This indicates the need for a closer look at specific aspects of a marriage rather than at family status as such. Marriage appears to pose a particular set of problems but also of advantages. Depending on a couple's choices, advantages or disadvantages may prevail. For instance, becoming a parent during the fellowship, as well as taking the fellowship to be with one's spouse, appeared to reflect less than optimal choices in terms of subsequent career outcomes.

The Collaboration Trap

Women's careers were less robust and resilient than men's when it came to the effects of some graduate and postdoctoral characteristics. There it seemed as if, for women graduate students and postdoctoral fellows, much more could go wrong—these relatively early career phases appeared strewn with more potential traps for women than for men. Many of these characteristics were in the area of interaction with other scientists—in the relationship with graduate and postdoctoral advisors and in general collaboration. This is precisely what one would expect according to Simmel's framework. If women graduate students are liable to be conceived as strangers who do not fully belong in their department (or in science in general) or are liable to act strangely in an unfamiliar environment, their interactions with mentors, advisors, and other scientists may be more fragile, more problematic.

It is interesting that women were slightly more collaborative than men in graduate school, but slightly less collaborative during the postdoctoral fellowship. This suggests that, as Reskin (1978) hypothesized, some male scientists found it easier to collaborate with women who were in a subordinate assistant role than with women who were in a more egalitarian collegial role. This hypothesis is also supported by clues that the nature of the advisor-fellow relationship was generally more problematic for women postdoctoral fellows.

In their subjective evaluations, women found the relationship with their postdoctoral mentors slightly less beneficial than the men did. The average rating of the advisors' career impact was "somewhat positive" for both women and men (4.05 versus 4.13 on a five-point rating scale ranging from 1 ("largely negative") to 5 ("largely positive"); $p = 0.4035$). But of the questionnaire respondents who answered open-ended questions about negative aspects of their postdoctoral fellowship, more than twice as many women (17.7 percent) as men (6.9 percent) complained about a negative impact of their advisors on their fellowship or subsequent professional careers (LRCS=9.4; p = 0.002). (Among all respondents, including those respondents who did not make any open-ended negative comments, the percentages of those who complained about their postdoctoral mentors were 12.0 percent for women and 3.0 percent for men (LRCS=19.5; $p < 0.0005$).)

Several women complained in open-ended comments in the initial questionnaire about having experienced collaboration in a subservient role. Some of the male mentors and senior collaborators may have had trouble entering an effective collaboration with a woman postdoctoral fellow. They may have tended to give her less demanding, more routine work and to treat her in a hierarchical master-apprentice mode as a perennial graduate student rather than as a colleague. Some women may also have been less insistent than men on getting their fair share out of a collaboration. Another possible response of mentors to women scientists was to ignore them more than men, which was evidenced in the slightly lower collaboration rate of female postdoctoral fellows and also in several comments of women respondents. In hindsight, some women considered their being ignored by advisors a blessing in disguise because it forced them to become independent researchers.

These sentiments of some women respondents about their collaboration with advisors were certainly consistent with our quantitative findings. For women postdoctoral fellows, as we have seen, collaboration with a scientist other than one's postdoctoral advisor presaged their subsequent employment in academe. Association with senior advisors adversely affected women's average productivity. Moreover, women respondents who closely collaborated with their advisor and did not have any other collaborators were found at a disadvantage in terms of later academic rank. It seemed almost as if some of these women were trapped in an intense subservient collaboration that had adverse effects on their later careers. This collaboration trap that tied women too closely to their advisors also illustrates that it is important for women who want to become academic scientists to form collegial networks at the postdoctoral stage.

The fellows' subjective evaluations of their postdoctoral fellowship further illuminate the issue of collegial networks. We noted that 74.6 percent of the men but only 65.5 percent of the women (LRCS; $p = 0.018$) agreed that the fellowship gave them more exposure to scientific stimulation in the form of symposia, colloquia, or opportunities to present papers than they would have had without the fellowship. Of those who did experience such contacts, the women rated the effects more positively than the men (4.12 versus 3.89; $p = 0.0227$).[6] Similarly, 78.2 percent of the men agreed with the statement that the fellowship had increased their opportunity to make professional contacts that led to post-fellowship support, but only 66.8 percent of the women did so ($p = 0.003$). Among those who did agree, the women again rated the effects of these contacts on their subsequent scientific careers slightly higher than the men (4.05 versus 3.92 on a five-point rating scale; $p = 0.1650$).

Together, these results form a consistent scenario: fewer women than men postdoctoral fellows had the opportunity of linking into collegial networks during their fellowship. And the women who had this opportunity rated its value higher than the men did. In sum, the connections we noted between the nature of postdoctoral collaboration and key career outcomes for women scientists were corroborated by the fellows' subjective evaluation of their collaboration with their advisors and of their postdoctoral networking opportunities.

An interesting issue in the area of collaboration is the advisor's gender. Here, too, a complex and partly counterintuitive picture emerged. Women who were associated with women postdoctoral advisors were more likely than those associated with men advisors to leave science later—a result that was entirely contrary to the role-model effect we had expected. Owing to the small numbers of former fellows with female advisors, one should, of course, not attach a great deal of confidence to this provocative finding. Nevertheless, we hesitate to dismiss it entirely as a statistical fluke, because some anecdotal evidence in our sample backs it up.

In our face-to-face interviews, several of the successful older women scientists who had succeeded in the face of a generally non-supportive environment held a gender-blind, strictly meritocratic view of the scientific system that prevented them from what they might have perceived as handing out special favors to younger women scientists. This may be connected to what has been described as "queen bee" attitudes of successful women (O'Leary 1988). A former Bunting fellow felt that "women who have reached some status in their field do not always help others, even though they like to get advertised as ones supportive of other women. In fact, my experience has been that men will support women more than women tend to support each other." Thus, while "students often expect women faculty to be 'easier' to work for—more forgiving, less demanding, etc.," as another woman respondent observed, these expectations may not always be justified. Moreover, role models need not necessarily be positive. If a young woman scientist found her female advisor's work or life situation undesirable, such a negative role model would actually provide her with an extra incentive for leaving science.

These findings suggest a generalization of Reskin's (1978) hypothesis that, as a woman respondent put it, "women are not gladly accepted by men as colleagues, as leaders in the scientific fields, or as major professors." It may not have been only male scientists who found it hard to interact with women as colleagues (that is, on an egalitarian basis) but scientists in general. In other words, there may not have been a well-established pattern of woman-to-woman collegial interaction either. The situation might, of course, have changed for the younger cohorts of women scientists who at present go through their postdoctoral phase. The available female advisors are now women who experienced the women's movement and might, on the average, have a stronger interest in remedying gender disparities in science.

Our main conclusion about graduate and postdoctoral experiences and their repercussions, which become evident both in the interrelations between domestic life and career and in scientific collaboration, is that it is very important to look at details. In terms of gender differences and disadvantages, this means that underneath a superficial similarity between the men's and women's graduate and postdoctoral experience there are subtle but important differences. For instance, if we just look at the number of postdoctoral mentors and their academic status, we find gender equality. If we focus on the nature of postdoctoral collaboration, however, we find that it tends

Table 3.13. Regression Models of Rank on Years Since Doctorate, Fellowship, Fields, Affiliation, Productivity, Graduate and Postgraduate Characteristics, and Gender

	1 Model without graduate and postdoctorate variables	2 Model with graduate and postdoctorate variables	3 Interaction model
Control variables			
Fellowship	.37‡	.31‡	.72†
Academic age	.08§	.07§	.06§
Biology	−.45‡	−.24•	−1.03†
PSME	−.21	−.14	.17
Affiliation	−.20•	−.18•	.14
Merit			
Productivity	.09§	.03	.00
Graduate variables			
Study time elapsed		−.09‡	−.05
Fellowship		.14	.12
Advisor's special status			.07
Collaborative research style			.15
Postdoctoral variables			
Qualifying function			.90•
No other job available		−.29†	−.29†
Took postdoctorate to be with spouse		−.18	−.12
Number of applications		−.14•	−.14•
Number of successful applications		.23†	.21†
Months as fellow		−.02§	−.01§
Junior advisor		.24•	.22•
Publication rate		.43‡	.43‡
Child born during postdoctorate		−.12	.53
Gender	−.27†	−.21†	.52
Interactions			
Control variables			
Gender × fellowship			−.29
Gender × academic age			.02

Table 3.13. *Continued*

Gender × biology			.60[†]
Gender × PSME			−.25
Gender × affiliation			−.26
Merit			
Gender × productivity			.02
Graduate variables			
Gender × advisor's special status			−.07
Gender × collaborative research style			−.12
Postdoctoral variables			
Gender × qualifying function			−.56*
Gender × child born during postdoctorate			−.56*
R^2	0.4971	0.6169	0.6613
N	281	281	281

NOTES: Unstandardized regression coefficients. Rank: 1 = nonprofessorial positions; 2 = assistant professor; 3 = associate professor; 4 = full professor. Fellowship: 0 = NRC; 1 = NSF. Affiliation: 0 = not at top 15 percent of graduate departments; 1 = at top 15 percent graduate departments. Productivity: number of publications per years since doctorate. Gender: 1 = male; 2 = female.
See table 2.2 for explanations of reference marks in this table.

to be fraught with more problems for women fellows, which express themselves in subsequent career disadvantages.

In chapter 2, we controlled academic rank—a key career outcome for academic scientists—for merit (publication productivity) to see if gender differences in rank achievement could be accounted for by differences in productivity. This approach was inspired by a process theory rather than an outcome theory of discrimination—we did not think that any gender disparity in outcomes was in itself already an indication of gender discrimination. As it turned out, we found that women respondents were disadvantaged in terms of academic rank compared with equally productive male counterparts. A lack of scientific merit was not sufficient as an explanation for the observed gender disparities, and the suspicion of gender discrimination was strengthened. But a sophisticated residualist would still argue that gender discrimination should only be assumed after all other relevant characteristics have been considered. Perhaps, women's later rank disadvantage could be accounted for by certain graduate or postdoctoral characteristics. Enter the gray and messy zone of human-capital variables to which this chapter has been devoted.

In table 3.13, we included the significant graduate and postdoctoral variables in the regression of rank. In the main-effect model (regression 2), gender remained significant. Even after controlling for relevant graduate and postdoctoral characteristics—in addition to merit and the usual control variables—women respondents, as a group, were still disadvantaged in terms of rank. This reinforces the explanation that gender discrimination may have caused this disparity. In the model containing gender interactions (regression 3), the importance of fields was again underscored. The gender disparities were concentrated in the fields outside biology.

In chapter 4, we will add another perspective on gender discrimination—our respondents' subjective experience of discrimination. We will then discuss women's subtle exclusion within the social system of science as well as their different style of doing science. This last aspect will also draw into doubt the suitability of publication productivity as a key indicator of scholarly merit—in the way that we, too, have used it so far.

4

CAREER PATHS AND CAREER OBSTACLES

IN THE NUMEROUS STATISTICAL analyses we carried out, we found gender disparities in career outcomes and various characteristics that could predict these outcomes (as well as other likely predictors that could not). This quantitative method produced a variety of interesting results, but it fell short of fully explaining the highly idiosyncratic career paths of our respondents. The statistical effect sizes were, although significant, mostly small. This was to be expected according to the theory of the accumulation of advantages and disadvantages. There was no single characteristic, no single choice that would have guaranteed certain success in a science career—or irrevocable failure. Science careers appeared to be shaped, to a considerable extent, by numerous idiosyncratic events and characteristics that are often insignificant by themselves but become forceful in their accumulation. For an examination of these subtle factors and mechanisms, we will now draw more heavily on our respondents' accounts of their career experiences, tuning in to the nuance and flavor of particular responses, often using their own words. This more qualitative approach will nonetheless be guided by quantitative findings in order to prevent purely impressionistic argumentation.

In terms of what the scientists in our sample actually do on the job, the genders were very similar. Contrary to the often-observed gender difference among academic scientists—that is, men tend to do more research and women tend to do more teaching (for example,

Chamberlain 1988, 263)—the time budgets of our women and men respondents were almost identical on the average, again reflecting the special population we selected for our research. The broad categories of a time budget do not, of course, reveal the details of the respondents' work conditions, and they say nothing about the career path that led up to the present position and the past and present obstacles the respondents had to face.

Indeed, in terms of career obstacles, female and male respondents painted different pictures. Women respondents as a group perceived their career path to be less smooth than men did. One manifestation of the lack of smoothness was, as we have seen, the much higher incidence of interruptions and part-time study during the graduate-school phase. Whereas 42.9 percent of the men in the large sample indicated that they encountered factors seriously interfering with their work as scientists, 61.5 percent of the women did ($p < 0.0005$). Among the women who noted any career obstacles in their questionnaire response, 12.0 percent had experienced outright discrimination—in other words, 7.7 percent of all women respondents. Among the small group of women who were no longer active research scientists, 21.4 percent referred to the experience of some form of discrimination as a career obstacle.

The women who described obstacles also mentioned being handicapped by bad interpersonal relationships with colleagues and mentors twice as often as men (10.3 percent versus 5.0 percent; $p = 0.079$). In addition, a notable gender difference was found in respect to family responsibilities. One-fifth (21.3 percent) of the women mentioned family demands, and 11.1 percent referred to their spouses' careers as impediments to their own careers. Very few of the men cited either of these obstacles (2.8 percent and 0.9 percent; $p < 0.0005$ in both cases).

When it came to the typical drawbacks in a scientist's career—the lack of financial and institutional support for research—the genders agreed: 39.3 percent of the women and 42.2 percent of the men mentioned obstacles of this kind. But half the men (50.5 percent) also complained about other activities that encroach on the time available for research, such as administrative work, committee duty, and teaching. Women, in contrast, mentioned competing activities less often (19.7 percent; $p < 0.0001$), perhaps because they were more inclined to regard teaching and service as part of their jobs rather than as obstacles.

The smaller sample of interviewees was also asked about particular advantages and opportunities that boosted their careers. Most respondents, men and women alike, mentioned material support (men: 58.2 percent; women: 52.9 percent; p = 0.46) and good luck and serendipity (men: 46.2 percent; women: 39.4 percent; p = 0.35). Twice as many women as men (14.4 percent versus 7.7 percent; p = 0.13) acknowledged social support systems as career advantages. Only 3.8 percent of the women mentioned Affirmative Action. Among advantageous internal traits for a researcher, more women than men mentioned resilience (81.3 percent versus 69.6 percent; p = .054) and the capability to work hard (84.8 percent versus 74.4 percent; p = 0.073).

Luck emerged as an important force in shaping career paths. As already noted, numerous respondents mentioned good luck and serendipity as career advantages. And in response to specific questions about luck in their careers, an overwhelming majority of interviewees acknowledged that good luck had affected their careers (men: 88.8 percent; women: 85.3 percent). Bad luck was acknowledged by a higher proportion of women than men (men: 33.7 percent; women: 49.0 percent; p = 0.033).

Luck in a science career can come in different shapes. First, creativity, that crucial element of scientific progress, eludes planning; and nobody knows from the outset if a creative idea will turn out to be useful. Both conceiving a creative hypothesis and having that hypothesis soon corroborated by experiments depend to some degree on luck. This type of good luck in actual research was, however, mentioned less frequently (men: 10.5 percent; women: 8.8 percent) than a second type—luck that takes the form of serendipity, as many interviewees called it (men: 90.8 percent; women: 87.5 percent).[1] These respondents acknowledged they were in the right place at the right time. Research programs and whole scientific fields may on short order turn hot or cold; promising ones may stagnate, while long shots may produce a breakthrough that catapults a researcher from the margins to the center of attention in the scientific community.

On the concrete level, serendipity was often described as meeting the right people. The right person can be a leading scholar who inspires the young researcher scientifically or a powerful figure in the science establishment who makes the right introductions and connections for the aspiring scientist. It can be someone whose personal integrity and kindness are impressive or someone who teaches the scientist how to play the quasipolitical game of furthering one's career in science. It can be someone who combines several of these aspects. But it should be noted that meeting the right people in one's field does not have to happen by sheer luck; it is one of the few externalities a scientist can make more probable by taking the initiative.

More women than men explicitly mentioned receiving fellowships among good-luck career events (18.8 percent versus 9.2 percent; $p = 0.086$). This appears to reflect the fact that, on the national average, women have received less fellowship support than men (Hornig 1987; Matyas 1991, 21). The women in our sample may have been aware that they were among the relatively few women afforded such opportunities, whereas a larger number of men may have taken their fellowships for granted.

Reverse serendipity—being in the wrong place at the wrong time—was the most common version of bad luck, although fewer people (men: 61.9 percent; women: 57.1 percent) reported it than mentioned serendipitous good luck. Bad luck in actual research was a somewhat more frequent response among the men (23.8 percent) than the women (7.1 percent; $p = 0.12$).

As we noted in chapter 1, other researchers have found females to be more likely than males to attribute their success to luck; and this attitude may be a liability for career success. Among our interview sample, however, there was no such gender difference in attribution; the vast majority of men and women acknowledge that luck—good or bad—was a decisive force in shaping their careers. This emphasis on luck may be due to the fact that science careers are indeed influenced to a particularly large degree by luck as compared with other professional fields. Science, by its very nature, probes into the unknown; and it is difficult to prearrange breakthrough discoveries or to position oneself on the particular part of the science frontier where the next major advance will occur.

We now turn to women's career obstacles in greater detail and, within our theoretical framework of deficit model versus difference model, discuss several major areas of potential obstacles for women:

gender discrimination and related phenomena (sexual harassment, sexism); informal exclusion; and general socialization differences. In addition, we address gender differences in doing science and in the normative concept of science. Finally, the impact of marriage and family is considered. It should go almost without saying that structural and internal obstacles—separated here for reasons of conceptual clarity—are intricately intertwined in the social reality of career paths.

Discrimination

Gender discrimination appears to be far from eradicated, according to reports from the total set of women whom we interviewed. When asked about sex discrimination, 72.8 percent of the women interviewees said that they had experienced such discrimination in their careers, while 12.9 percent of the men interviewees mentioned reverse discrimination. When asked about incidents that made them feel uncomfortable or surprised, 40.4 percent of the women interviewees mentioned episodes of sex discrimination, sexism, or sexual harassment. Thus, although outright gender discrimination may, in general, be on the decline, a woman entering science now should still be aware of the possibility that at some point in her career she might encounter behavior that she will consider gender discrimination.

Serious incidents of discrimination reported in a few cases by our interviewees included denials of jobs, promotions, or tenure when the women felt they were well qualified for a positive decision. One woman scientist related that her university "nominated me for the Non-tenured Women's Faculty Fellowship as the outstanding female researcher who does not have tenure, and, less than a year later, denied me tenure on the basis that my research was terrible. No, there is no logic to that, but it happened." Another woman respondent reported that she was denied a job because the hiring committee felt they should protect her marriage.

Such serious incidents of discrimination were less frequent than cases of insignificant slights. But one-fifth of the women interviewees (21.5 percent) said that discrimination had been a career

CAREER PATHS AND CAREER OBSTACLES

obstacle: 16.8 percent of the women interviewees mentioned discrimination by superiors, and 4.7 percent noted discrimination by their peers.[2] As one might expect, women who had left science were much more likely (38.5 percent) than academic scientists (19.4 percent) and women scientists outside academe (18.8 percent) to mention discrimination as a career obstacle. A few male respondents reported that reverse discrimination had hampered their careers. A man, for instance, noted, "I am all for affirmative action for blacks and other minorities, but not for women. Being a white male is one reason why I am at a teaching college despite over 30 publications and $180,000 in grant money."

Because gender discrimination is now illegal and violations can result in costly litigation and fines for the institution, women are never told that they are being rejected because of their gender, if that is the reason. Thus, women are apt to try guessing at the "real" motives behind a negative decision. This situation is very different from past days when gender discrimination was much more overt. A woman scientist in her late fifties described an incident that happened in the 1960s, when she had two small children. "I had been promised a fellowship in the department of medicine at [X University], and I hired a housekeeper and got everything all together in my household. After a month, I went and presented myself to the man who had offered me the fellowship. And he looked at me and said: 'I have no place for housewives in my department.' It was quite a performance—the kind of thing that would never happen in this day and age." Several older women, acknowledging that sweeping changes had occurred in the situation of women in the sciences, expressed regret that they were not born later.

Sexual harassment was reported by 11.7 percent of the women interviewees. The following is a classic example. The thesis work of one woman graduate student was evaluated by a renowned expert outside her own university. When they finally met face to face at a conference, he initiated an affair with her. "Although I participated fully, I think it was a real abuse of power, and it was unethical of him." One of the more bizarre cases of sexual harassment was told by another woman scientist who works in academe. "The day I was to be voted on for tenure here, one of the associate professors walked into my office at 10 o'clock in the morning and said, 'Either you go to bed with me at noon, or I'm voting No for you at 4.'" She added that she "threw him out" and received tenure nonetheless.

It is important to keep in mind that in less self-evident cases the same event may be perceived as discriminatory by one person and not by someone else. Two patterns appeared to shape the women respondents' perception of discrimination. One was a societal trend, the increasing awareness of gender discrimination and harassment. The other pattern linked individual career outcomes to the awareness of discrimination.

Several women reported that in hindsight they considered some experiences discriminatory but did not judge them to be so when they happened. A female respondent said, "I don't think I could have described well enough what I was experiencing when it was happening, to get help with it." She credited the women's movement with clearly articulating, or "naming," experiences of discrimination that previously were not acknowledged as such. In these cases, the women's movement provided the conceptual tools for reinterpreting an experience as discriminatory.

When considering the interviews as a whole, two opposite types of stories seemed to emerge. In one, discriminations were minimized; in the other, they were maximized. In the first type of story, the far more frequent type, hardly any mention of discrimination was volunteered. Details came to light only on demand, in response to a specific question about gender discrimination. The other type was an account prominently featuring a protracted list of discriminations. Women whose careers floundered were more likely to relate accounts that contained a whole series of mistreatments. It seems all too easy to dismiss these stories as psychological rationalization—in the sense that a career failure may provoke feelings of bitterness and anger and heighten a woman's perceptiveness of discriminatory episodes. The predominance of discrimination in these career stories may well indicate a sociological process—predicted by the theory of the accumulation of advantages and disadvantages—in which discrimination is likely to beget further discrimination. For instance, if a woman unsuccessfully fights a discriminatory tenure denial, she might gain the reputation of a troublemaker that may then lead to the denial of other employment opportunities.

Many women, particularly the reasonably successful ones, appeared to employ a very pragmatic strategy for dealing with discriminatory incidents of the less serious kind that appear to be pervasive. These women carefully chose their battles and navigated around the less severe types of discrimination rather than confronted them head on. The following elements of this strategy emerged.

Ignoring

During their social interactions with other scientists, women are bound to encounter, in the words of a woman respondent, "50-year-old men who don't know how to talk to you, or call you 'sweetie' or 'honey' . . . But I get around it, or I just ignore it. No, they shouldn't treat me any different than their male counterparts, but I mean, what are you going to do, make an issue of it, or just go on and get what you want and be done with it? And that's the way I've approached things." Several women counted a certain thick-skinnedness among their advantages.

Humor

Some women described their approach as trying to defuse and circumvent potential discrimination by reacting in a cordial way. This cordial, even humorous, approach was regarded as more productive than a confrontational one. "It is clear to me there are difficulties, and there are certainly prejudiced men. However, I find it is easier, and I think one is more effective, if one finds a nice, joking way of conveying that rather than being hostile."

Compliance

In some instances, women just complied with what they considered trivial requests rather than started a fight about points of principle. For example, the male supervisor of a woman scientist in industry wanted her to change her appearance to look more like a management consultant, which meant wearing suits and pinning her pigtail up. Although she would have probably had some grounds to protest this request, she complied because "if you can make your boss happy with four oversize bobby pins, you're an idiot not to do it."

De-emphasizing Gender

In an effort to avoid gender-related problems, some women said they de-emphasized their femininity and emphasized their professional identity as a scientist. As one woman put it, "I think to make it as a female scientist, you have to be a scientist and truly downplay the female." But downplaying one's femininity is, of course, particularly difficult for pregnant women. In the words of another woman, "being a pregnant woman is the most awful of all. You're not treated with the deference that a cute young thing is treated with. And never are you more loudly proclaiming your womanliness, so the comments get made about the seriousness of your pursuits. Everybody thinks it's fair game to ask an obviously pregnant woman exactly what her plans are, or things like that."

Avoidance

One woman said that she had been able to detect potentially discriminatory situations from the outset and then immediately avoided them. "The times that I sensed that there could be a problem [with gender discrimination], I was always in control from the standpoint of finding a new advisor or getting allegiance with someone else who didn't feel that way. I never stayed in a situation that I thought was going to be detrimental. I always got out of it at the first hint."

Informal Structural Obstacles

Our respondents also talked about the more informal obstacles to women's science careers. Many women reported being excluded from informal social events with their colleagues, such as playing sports together or going out for drinks. In the old days, women might have been explicitly told not to participate in such gatherings, as one woman scientist reported. More recently, explicit exclusion has become rarer, but there remain some cultural factors that make women feel out of place in a predominately male group of colleagues. One woman interviewee noticed, "there's always a sense, especially in a group that does not include many women, that you're not one of the guys, and that works against you, and that is impossible to fight, of course." At this level of subtle and informal gender segregation, the deficit and difference models converge. Informal exclusion of women by men can have its complement in women's reluctance to participate in types of social activities that do not appeal to them.

Social interactions with colleagues may affect scientific careers. About half of the interviewees, men and women alike, thought that their social interactions with their peers had an impact on the progress of their career (men: 51.1 percent; women: 49.5 percent). Some women felt that their exclusion from informal contacts negatively influenced their career; important information is being exchanged and decisions are often made at casual meetings among colleagues, and women's absence at these occasions may render them invisible. As one woman poignantly put it, "certain things were clearly closed to me. There were little social hours and stuff that involved male camaraderie, that I really didn't feel like busting in on. But I knew that those were the places where hiring decisions got worked on."

Even if bans on outright discrimination are energetically enforced, the effects of unofficial decision making (which is much more difficult to police) may ensure the same end as surely. For example, if leading male scientists think that an inborn factor makes women less capable in certain fields of science, they may be inclined to ignore their female colleagues and cut them off from the flow of insider information. Ultimately, these attitudes might function as a self-fulfilling prophecy.

Double Standard

Some women also reiterated the often-noted double standard for male and female behavior: behavior that is acceptable in men may be unacceptable in women. For instance, an assertive woman is likely to be seen as bitchy, whereas a man with the same temperament might not be so judged. In a vivid description of the double standard, a woman interviewee talked about a male scientist who did a postdoctoral fellowship simultaneously with her in the same research group. This man "was always running around screaming, and very temperamental and just impossible. A woman could never have gotten away with acting like that. We'd always have to be more conciliatory, more mainstream, more normal. And this person was absolutely paranormal, but he was viewed as a brilliant scientist." In this respect, it is interesting that, when asked about gender differences among scientists in general, a few men interviewees (5.6 percent)—and none of the women—thought women scientists were *more* aggressive, competitive, or risk-taking, possibly because these men adhered to a traditional standard of female pliability. By contrast, 12.4 percent of the men interviewees said that women had fewer of these traits, as did 9.8 percent of the women.

Collaboration and Integration into The Social Network of Science

A look at scientific collaboration should elucidate how subtle structural obstacles might marginalize some women scientists. We thought a lower proportion of collaborative publications might indicate women's relative isolation. But we found no substantial gender difference in the proportion of collaborative publications among all publications for our academic respondents (excluding mathematicians), replicating a finding by Cole and Zuckerman (1984). We should note here, however, that in many cases the alternative to a collaborative publication may be no publication at all rather than a solo publication. Thus, a relative marginalization of women scientists might show up in the lower rate of publication productivity (described earlier)—not in a lower proportion of collaborative publications.

Another aspect of collaboration is membership in research teams. The same proportion of women as men scientists in our sample (excluding mathematicians) were members of research teams (men: 67.3 percent; women: 68.5 percent; academic men: 61.4 percent; academic women: 62.1 percent); but the average team size was somewhat, albeit not significantly, larger for men (men: 14.1; women: 10.2; $p = 0.4697$; academic men: 11.6; academic women: 8.8; $p = 0.1975$).

Furthermore, similar proportions of men and women academic scientists noted that they had supervised at least one graduate student's Ph.D. thesis (men: 65.1 percent; women 62.8 percent). Women and men academics also had identical output rates of graduate advisees per year once they had started being supervisors (men: 0.66; women: 0.66). But although no gender difference existed in the formal advising of graduate students, women did have significantly fewer graduate assistants in their research teams (men: 3.99; women: 2.42; $p = 0.0147$). Women were thus handicapped in terms of collaborating with graduate research assistants whom they were not formally advising. Men appeared to be better able to attract these "free-floating" graduate students into their research teams, which may have given them an advantage in terms of research productivity.

Thus, at a glance, gender similarities in collaboration outweighed gender differences. Nevertheless, gender differences may lie, as Fox emphasized, in the fine structure of interaction: even if the previously noted characteristics are similar for men and women, "the nature and quality of collegial interaction may be different" (Fox and Ferri 1992, 267; compare Fox 1991). On the one hand, some small details, such as the relative dearth of collaborating graduate students, pointed to a slight marginalization of the women in our sample. On the other hand, the qualitative material provided additional indications of the slight distancing of women from the collegial information exchange.

Our women respondents were certainly conscious of subtle effects of this kind. More women than men thought they had, at least sometimes, interacted differently with male and female colleagues (men: 40.4 percent; women: 54.4 percent; $p = 0.055$). As a female scientist observed, men "talk to each other in a way that is slightly different, in terms of sharing problems and stuff like that, from what one's able to do if you're of the opposite sex, and if you don't have people to talk to these things about, you have to be fiercely independent." Another woman concurred: "Men are much more open with each other in their research, willing to talk about their work together 'over a beer,' and collaborate more. Women are not so tight-knit, and tend to work harder alone to compete (in number of publications, etc.) with men,

Table 4.1. Collaborative Research Style at Different Career Stages by Gender

	Before fellowship	*During fellowship*	*After fellowship*
Women	2.18	2.33	2.68
N	188	189	183
Men	2.10	2.56	2.89
N	498	498	490

NOTES: The numbers are averages on a five-point rating scale, ranging from 1 ("I usually worked alone on my scientific problems") to 5 ("my research was performed almost exclusively in collaboration with other scientists").

in an environment created for men." When asked whether their gender influenced their working style as a scientist, more women than men said they liked working alone (men: 6.0 percent; women: 16.2 percent; $p = 0.060$).

Especially for the later stages of their science careers, our women respondents reported a marginally lower degree of collaboration. Whereas before the fellowship the women were slightly more collaborative than the men, the women collaborated less than the men both during and after the postdoctoral period (table 4.1). The aggregate data trend for the men toward more collaboration over the three career stages (graduate school, postdoctoral fellowship, now) was replicated at the individual level. Only 27.7 percent of the men adopted a less collaborative style either between graduate school and postdoctoral fellowship or between fellowship and now. By contrast, almost half the women (47.1 percent) became less collaborative over time.

Thus, women as a group experienced less of what is usually considered a mature type of collegial collaboration with peers or collaboration with their own students (during the postdoctoral fellowship and later), whereas they experienced more collaboration as junior partners (during graduate school). Men may have had an easier transition during the postdoctoral phase into collegial collaboration. Recall that collaboration during the fellowship had opposite effects on later career outcomes for men and women. A number of highly collaborative female fellows may have experienced detrimental effects of the previously described "collaboration trap." A look at the research style by field supports this observation. In biology—where women enjoyed relative equality in career outcomes—the women's average degree of collaboration closely tracked the men's average in all three stages (before, during, and after the fellowship). In PSME, by contrast—where women faced considerable outcome disparities—women had a dramatically higher intensity of collaboration in graduate school than their male cohorts did (men: 2.09; women: 2.58; $p = 0.0109$); and this gender gap closed in subsequent stages.

Women may face informal structural obstacles to their full participation in scientific collaborations when some men shy away from collaborating with them. And additional obstacles may lie within the structure of collaboration so that women themselves avoid collaborations because they have experienced or are anticipating trouble in joint efforts. On the one hand, as long as women are considered strangers in science, the burden of proof for their competence is reversed. As a woman respondent observed, "men with a Ph.D. are taken for granted to be knowledgeable and professional until proven otherwise. It is unfortunately the other way around for women." This often heightens critical attention to a woman's scientific work. In the words of another woman, "men can work on a problem with less scrutiny than women, who are often 'under the magnifying glass.'" Some men may thus hesitate to collaborate with women to the same degree as with their regular (that is, male) colleagues.

On the other hand, some women said they avoided collaborations out of fear that their work would be misappropriated by collaborators. One woman, for instance, expressed the "sense that things would be taken away from me, if working with others." Often, this fear was based on bad exeriences with collaboration in the past, when women had not received the credit they felt they deserved. The dearth of women's first-authored collaborative publications would indicate that women, as a group, tended to play subordinate roles in collaborations.[3] In our sample, the proportion of first-authored publications among coauthored publications was slightly lower for the women, although the gender difference did not quite reach significance level after controlling for fields and annual productivity rate. Cole and Zuckerman (1984) also found no such gender difference. Nonetheless, the fear of becoming or being perceived as the junior partner in a collaboration may be a contributing element to women's slightly less collaborative research style in later career stages. A woman interviewee, for instance, noted:

> I was very much an isolationist, in terms of my work. And it's a habit that's kind of hard to break, and I had a very specific reason for being that way: there were no women to collaborate with anyway, and my very strong feeling was that any work a woman did as a coauthor or a co-principal investigator was always credited to the male, and I damn well wanted credit for what I was doing. So I wasn't going to dilute that by letting five guys have their names on what I was doing, and it was a tough decision to maintain, and it's probably the only reason I survived as long as I did as a recognizable researcher in the field.

Because some women perceived heightened risks in collaborating with male colleagues, they found collaborations with other women preferable. With increasing numbers of women scientists, collaborations among women may become more frequent.

Finally, personal sociability does not sufficiently explain the degree of a scientist's marginalization. Even men who regarded themselves as nonsociable successfully operated within the collegial network, whereas women scientists who regarded themselves as social persons remained somehow isolated. A very successful female scientist said she enjoyed the social aspects of doing science with her collaborators in the laboratory, and another successful woman also thought she was good at relating with people; but they both reported some isolation from the collegial network. A successful male scientist, by contrast, although confessing to a certain lack of social skills, still managed to make contacts and connections with other scientists that proved crucial for his career success.

Sexual Dimension

The sexual dimension can be a complicating factor in collegial relationships. Of the men interviewees who said they interacted differently with men and women colleagues, 20.6 percent referred to sexual tension in interactions with colleagues of the opposite sex as compared with 7.3 percent of the women ($p = 0.097$). Mindful of potential problems in this area, a woman scientist in academe recounted how she curtailed her social interactions with colleagues to avoid any possibility of sexual misinterpretation.

The "awkwardness of close interaction between men and women" was also noted by another woman respondent who had been "hesitant to pursue collaboration if there was any possibility of my motive being interpreted as not entirely scientific. I continue to actively avoid collaboration with men that are attractive and charming." This awkwardness was also sometimes experienced in the interaction between senior females and junior males. A female former postdoctoral fellow, who now advises postdoctoral fellows of her own, noted, "I still must make a conscious effort to be at ease in taking my male postdocs out to lunch or inviting them for a beer after the seminar. Since so much satisfaction in the early years is based on social interactions, I'm sure this affects their productivity and thus mine." Another female scientist said that it is just impossible for a woman scientist at a conference to invite a male colleague to her room for an informal discussion about research over drinks, whereas this is very common among male scientists. Some women scientists thus appeared to experience a clash between the intrascience norm of seeking out contacts with professional peers and the traditional cultural pattern that decent women do not approach men.

The Pressure of Being a Good Citizen

Women sometimes feel particularly high pressure to take on administrative duties. If there are only few women in a department or a division, these women appear to be in high demand for various committee duties so as to include a woman's point of view. A woman respondent noted, "I am regarded as a token woman scientist and asked to be a woman representative on many school affairs because of this. My fellow men scientists seem to get more research done because they are not asked to be representatives on school affairs to the extent that I have been asked." Typically, women have a hard time declining these tasks. As another woman respondent observed, women "are often asked to do additional duties which take time away from their research. Often they accept these tasks—in fact, if one is untenured it is usually a bad idea to say no—although most accept without thinking of saying no."

Socialization Differences

Even our group of atypical women who became doctoral-level scientists differed from their male cohorts in their own estimation of self-confidence, ambition, and related traits. These differences are consistent with the traditional gender pattern of socialization, although we should also note that they may, in part, be the result of women's experiences within the social system of science. Substantially more men than women among our interviewees reported that they considered their scientific ability to be above average (men: 69.7 percent; women: 51.5 percent; $p = 0.011$), whereas more women than men considered theirs to be average (men: 18.0 percent; women: 34.7 percent; $p = 0.009$). Similar results were found when the interviewees were asked to evaluate their technical skills. In addition, more men than women interviewees thought others rated their scientific ability above average (men: 70.0 percent; women: 52.6 percent; $p = 0.018$); slightly more women than men thought it was rated as average (men: 17.5 percent; women: 23.7 percent; $p = 0.31$).

Slightly more men than women interviewees considered themselves self-confident (57.6 percent versus 48.6 percent; $p = 0.21$). And when asked whether they should have handled their career obstacles in a different way, many more women than men thought they should have had more confidence or been more assertive (25.3 percent versus 4.6 percent; $p = 0.0001$). In our interview sample, women were on the whole more likely than men to have entered their science career gingerly, taking a step-by-step approach rather than having clear overall career goals at the outset. More than three times as many women as men (15.9 percent versus 4.4 percent; $p = 0.006$) said they had vague or unclear career aspirations when they started out in science. Even now, only 34.3 percent of the women, as compared with 44.2 percent of the men, said they were more ambitious than the average colleague ($p = 0.173$).

In a difference-model approach, one would consider such gender differences among the causes that make women scientists, on average, less successful than men. Indeed, within the group of interviewees currently working in academe, the mentioned traits were commonly related to our composite measure of success (a weighted average of publication productivity, academic rank, and prestige of affiliation). The interviewees' ambition ($r = 0.34$; $p = 0.0015$), their evaluation of their own scientific ability ($r = 0.36$; $p = 0.0009$), as well as their estimation of how their colleagues would rate their scientific ability ($r = 0.24$; $p = 0.0444$) correlated considerably with actual attainment of success in academe. A weaker correlation was found for self-confidence ($r = 0.16$; $p = 0.122$).

We should emphasize, however, that these correlations do not indicate a direction of causality. Ambitious scientists, for instance, may become more successful; or successful scientists may become more ambitious. Thus, rather than being the cause of the women's average lag in career success, the gender differences in the noted traits might be the effect of women's greater career obstacles and ensuing relative lack of success—a view that proponents of the deficit model might take. Again, there may be some reinforcing combination of both processes that intricately links elements of the difference and the deficit models in real career paths in science.

When asked about internal obstacles, more women than men interviewees mentioned their personality (59.6 percent versus 42.0 percent; $p = 0.0164$), but more men than women referred to bad work habits (21.6 percent versus 7.1 percent; $p = 0.0052$). Apparently, the men tended to fault specifics, whereas the women tended to blame more global features of their personality—making it harder for women than for men to address a particular problem that may hold them back and to initiate change.

In the pursuit of their careers, men appeared to be more independent of their social surroundings. As a woman scientist put it, "male scientists are very internally directed, and they don't listen to what people are saying around them." Twice as many men as women asserted in our interviews that nobody helped them deal with their obstacles (20.7 percent versus 9.1 percent; $p = 0.029$). By contrast, more women than men acknowledged family and significant others (62.6 percent versus 42.5 percent; $p = 0.006$), colleagues (49.5 percent versus 34.5 percent; $p = 0.039$), and friends (31.3 percent versus 14.9 percent; $p = 0.008$) as sources of support. Thus, women scientists evidenced a relatively great need for support that stemmed from two sources: As we mentioned earlier, more women than men respondents reported that they had encountered obstacles on their career paths; and, as we have just seen, women seemed to be more dependent on social support systems in overcoming obstacles. This latter result corresponds with theories that propose a greater social connectedness of women, which, of course, poses the question of whether women receive this much-needed support within the current system of science.

We also found other evidence of the existence of—and the need for—greater connectedness among women. In her interactions with her co-workers, a female academic scientist observed that "the currency of conversation tends to be slightly different" from how her male colleagues interact with their co-workers; that is, she preferred a more personal interaction style. Similarly, another woman said that the members of her graduate-student research groups were more on "family terms" than was usual for male-led groups. The family imagery was expressed even more vividly by a third woman scientist: "I've tried to create families out of my collaborations with people. The relationship is equally important to me as the work itself. I pick people that I like working with, and then we do projects that I consider to be really fun."

By themselves, these few statements may not suffice to postulate a general trend among women scientists to consider family relations the prototype of human relations to be emulated in the science domain. They do, however, exemplify what has been termed the "historical allegiance of women to the family model as an ideal form of organization" (Nowotny 1991, 155). Residues of this allegiance may partly explain a certain uneasiness some women felt with the common type of scientific collaboration, which is more fleeting and superficial than they might prefer (even when, as is fashionable, the word *family* is used by groups of collaborators).[4] Because these women scientists make heavier emotional investments in their collaborative relationships, they might choose fewer and closer collaborations compared with men's collaboration patterns. This female collaboration style may contribute to the marginalization of some women in the social system of science. One of the most successful female academics in our sample, for instance, noted that the trend in her field to do research in larger and larger collaborations might restrict her in the future because she is "not a very good joiner" and prefers to work with a small number of other people and her students.

Family-style collaborative relationships may be more rewarding, but they can also be more disastrous because they intensify both positive and negative emotions toward colleagues. After all, if the failure of a collaboration were also seen in family terms, it would be the equivalent of divorce, bereavement, or abandonment. A broken family causes more problems than disbanding what is viewed as a temporary business arrangement.

Styles of Doing Science

Gender differences were thought to exist also in the ways of doing science. Somewhat more women than men interviewees believed in the existence of gender differences in the work of scientists in general (men: 49.4 percent; women: 60.8 percent; $p = 0.117$). Substantially more women than men also thought that their own gender influenced the way they did science. The area of science in which such a gender influence was most commonly thought to play a role was professional conduct (men: 25.6 percent; women: 51.2 percent; $p = 0.0006$).

Fewer respondents thought there were gender influences on the choice of their research subject (men: 15.7 percent; women: 40.0 percent; $p = 0.0004$) and on their way of thinking in science (men: 20.0 percent; women: 36.0 percent; $p = 0.019$). The lowest proportions were found in respect to methods adopted (men: 9.9 percent; women: 34.8 percent; $p = 0.0001$). In all these respects, women respondents appeared to be much more convinced than men that gender influences the way of doing science, whereas men appeared more prone to associate themselves with the traditional gender-neutral doctrine of science. Of course, we are dealing here only with scientists' perceptions and self-reports, which are not necessarily based in reality (that is, in measurable gender differences in actual scientific behavior). Nonetheless, if, in the minds of a sizable proportion of scientists, gender has become a relevant variable for interpreting scientific behavior, this fact in itself is bound to have some repercussions on the notion of gender-blind science.

How are these beliefs about gender differences related to career outcomes in science? A plausible assumption might be that successful women scientists are less inclined than undistinguished ones to think that their gender influences their way of doing science. A male respondent observed that "the best female scientists work on the same things in more or less the same way men do." And after all, as a woman who did not believe in gender differences in scientific style put it, "to succeed you must play the game the way men play it."

Nevertheless, we found our outcome measures (leaving science, area of science employment, composite of academic success) to be generally unrelated to the various aspects of the belief in gender differences in science (both among men and women). There were a few exceptions. Academic scientists were marginally more likely than nonacademic ones to report an impact of their gender on their way of thinking (32.7 percent versus 19.2 percent; $p = 0.070$), and they also believed to a larger extent in gender differences among scientists in general (65.2 percent versus 42.4 percent; $p = 0.004$). The greater proximity in a university setting to current strands of sociological and feminist theories that postulate the importance of gender may have made academic scientists more aware of the gender dimension.

The following sections concentrate on the perceived gender differences in scientific style, but we should not forget that a large majority of the male interviewees and a majority of females in some respects (and a substantial minority in others) did not believe that their gender influenced their way of doing science.

Women's Professional Conduct: Less Careerist

In general, professional age was unrelated to the belief in gender differences; but younger women interviewees were more likely than older women to believe that their gender influenced their professional conduct. This may indicate an increasing awareness and sensitivity to the gender dimension in the interactions among scientists rather than an actual divergence in male and female interaction styles.

In the larger sample, the women who responded affirmatively to an open-ended question about gender differences among scientists most frequently mentioned that women were less aggressive, confident, or took fewer risks (women: 37.8 percent; men: 23.7 percent; $p = 0.027$). Such gender differences in personal predispositions and interaction styles may have a considerable impact on men's and women's career progress in science. A common observation among the interviewees, both men and women, was that men scientists have more "entrepreneurial spunk," as a female interviewee called it. They are, in this view, more aggressive, combative, and self-promoting in their pursuit of career success and so they achieve higher visibility—in short, they are better at playing the political game of career advancement. Such a gender difference in careerist fervor may, to some degree, also be connected with the fact that men are much more likely than women to be primary wage earners. Among our married interviewees, 80.3 percent of the men but only 34.2 percent of the women said they were the primary wage earners ($p < 0.0001$). Differences in professional conduct, although noted here among the difference-model obstacles, may of course be structurally reinforced and then also take on the character of deficit-model obstacles.

Furthermore, some women interviewees reported that men have a way of showing off at conferences that alienates women. The following comment by a female scientist illustrates this male behavior as well as the relatively greater inability of females to initiate contact with males.

A lot of the connections that people make at meetings and so on, I couldn't do, because what men did is they stood in the hallways and found the great men and went over and shook their hands or asked them to have a drink with them or something, and women couldn't do that in my day. So you couldn't initiate anything . . . but if you were in a group of people who managed to connect with that person, then you could try to talk to him. But more often than not, when you were in a group like that, the men showed off for each other, they took themselves terribly seriously and they said any kind of thing that came to their head. I call it "professor talk," and I can do it very well. To me it's a big joke, but it's something that men do all the time with each other, and it's a kind of stroking behavior, makes them all feel good, even though they're really not talking about anything and they're really not communicating very much. And I found that a waste of my time.

Although professor talk may indeed be a waste of time in terms of exchanging research information or gaining scientific insights, it may be anything but wasteful in terms of its hidden agenda. Professor talk—or a "bull session" or "chatty self-promotion," as other women respondents called it—appeared to have the important social subtext of a bonding ritual. And the social bonds thus forged may indeed have some beneficial impact on a scientist's research and career later on. This again reflects gender differences in interaction styles. Men's greater independence in their career pursuits, which we previously noted, finds its complement in what some women might consider superficial and hollow rituals of establishing social contact with colleagues.

Individuals set their own career goals and define what success means for them. Some of the women interviewees were certainly as ambitious if not more ambitious in terms of career success than many men. But we already noted a slight gender difference in ambition, on the average. The typical woman among our interviewees set her goals somewhat lower than the typical man did, which may be connected with women's less aggressive conduct.

Judging by their responses, few of our scientists, male or female, were hungry for scientific power. A large majority of both men and women interviewees appeared to be burdened rather than elated by their opportunities to make decisions affecting other scientists' careers, and nobody confessed to relishing administrative tasks. Women scientists, however, reported themselves to be even less keen on influence and power than their male counterparts. In the words of a woman respondent, women scientists are "more interested in the science (as are many men) than in 'getting to the top.'" An eminent woman scientist thought it typically male to aspire to chairmanships or deanships, and she said she did not harbor such aspirations at all.

Most of the successful women respondents could be described as reluctant leaders. They recognized that their elevated position in their fields had given them a substantial amount of power—some of them seemed to think perhaps too much power—and they intended to be very scrupulous in exercising this power fairly. By their accounts, these women scientists had either fulfilled most of their aspirations already or their yet-unfulfilled ambitions were science internal, such as the recognition of being elected to the National Academy of Sciences.

Whereas most of the men had a similar attitude, a few successful men scientists were quite comfortable with the power aspect of their position. With great ease, one of them acknowledged that he was a leader in his field with a large amount of power. And another male respondent, who had served a term as departmental chairman, acutely appreciated the division of labor in science. He explicitly defined his role as directing a large-scale research operation. In sum, our women scientists appeared to be the "purer" scientists—they were somewhat less concerned with the political aspects of science careers, such as influence and power, than their male counterparts.

Except for a small number who were disaffected with the culture and social conditions of scientific labor, most of our interviewees expressed their strong enthusiasm for doing research. Many felt privileged that they had the opportunity to do in their jobs what they enjoyed most. As a male scientist put it, "so much fun, and you get paid for it, too." A large majority of the interviewees (men: 80.2 percent; women: 77.9 percent) said they were satisfied with their life in general. The most popular responses to a question about the favorite aspect of doing science referred to the actual process of problem solving in scientific research (men: 52.3 percent; women: 60.0 percent; $p = 0.295$) and to the gaining of new knowledge and insight (men: 58.1 percent; women: 50.0 percent; $p = 0.269$). Although the gender differences on these two responses were not significant, they might indicate, when viewed together, that the men were somewhat more interested in the end result of research than in the research process itself. The women, by contrast, slightly emphasized the intellectually stimulating process of scientific research over the bottom line of results: a new research finding and a new publication. This, again, might signify a less careerist and strategic approach to research that might slow women's careers at various junctures when scientists' track records are assessed.

Women's Problem Selection: Niche Approach

In respect to choosing subfields and problems, a number of male and female respondents agreed with a woman who noticed "fewer women in highly theoretical/mathematical sub-fields (including myself—I am not much of a mathematician, compared with many of my male colleagues)." But gender differences appeared to go beyond differences in mathematics interest or training. In terms of problem selection, many women reported to be following what could be termed a niche approach—creating their own area of research expertise. One respondent observed that "women may shy away from very competitive projects more than their male counterparts." A good example is a female interviewee who liked "to sense that I had my own area, that I wasn't just a cog." Similarly, a woman respondent said that she was predisposed to selecting research problems that were completely her own because "I very much dislike working on problems that I know other people are working on." Rather than competing with other researchers and research groups in a race toward the solution of the same problem, she carves out a niche for herself. Another woman scientist, who said she did not follow a niche approach, nonetheless saw the phenomenon to be widespread among her female colleagues.

A female respondent observed: "Although men and women may do similar research, women tend toward long-term problems which involve a lot of detail, while men want to do dramatic things which bring grand results quickly." Men seemed more likely to join the fray in hot fields and breakthrough topics. A male scientist, for instance, said he consciously seeks out problems that at the moment command a great deal of interest in the scientific community. Another man observed that "as a male, it's easier to pick a more ambitious topic, say you're a graduate student and you're working on a doctoral dissertation." According to this scientist, men are better able to be "maverickish or nonconformist," whereas women stick more to the established rules and procedures. Although these mavericks operate somewhat outside established science, they often do not want just a quiet niche for themselves but aspire to bring about major changes in science—moving their field to the forefront. A male interviewee typifies such a maverick. From early on in his career, he was among the founders of a new field of scientific research despite some hostility from scientists in the neighboring traditional disciplines.

Women's niche approach may also be reflected in the focus of their professional identity. Fewer women than men interviewees identified themselves by the broader label of "researcher," "scientist," or "academic" (men: 59.2 percent; women: 44.4 percent; $p = 0.065$), but twice as many women as men focused their professional identity on a subdiscipline (men: 9.2 percent; women: 19.8 percent; $p = 0.060$).

Impressions of women scientists' niche approach were widespread among our respondents, but we should note once again that there are, of course, conspicuous examples to the contrary—both women scientists who pick the hottest topics and men scientists who prefer a quieter niche. Gender differences in problem choice may be decreasing, at least in the perception of our scientists. Younger interviewees were less likely than older interviewees to believe that their gender influenced their choice of research subjects.

Women's Methodology: Perfectionism

When a specifically female methodological approach and way of thinking was remarked on by our interviewees, it usually did not conform to some alternative approach, such as a nonandrocentric science occasionally discussed in the literature (see chapter 1). Among the rare exceptions was a woman geologist who described herself as having an increasing feeling of "oneness" with a mountain the more she analyzes it. And another woman noted, "There very likely is a difference in the degree of holistic vs. logical/sequential thinking that is innate."

The overriding theme that emerged from the interview responses about gender differences in the methodological approach was that women were said to be more *cautious and careful in their methods* and to pay more attention to details. A woman respondent stated that "women are often more careful in their research and more hesitant to make statements until they feel they can really 'prove' them." In the words of another woman respondent, "women are more thorough, less likely to shoot from the hip." Yet another female respondent observed that women tend "to wait until some piece of work is 'perfect' before we publicize it." Numerous women acknowledged a tendency to be perfectionists in their scientific work. Some of them said they were perfectionists because they wanted to avoid failure or criticism. Another woman scientist stressed women's attention to detail: "Women are more meticulous; they tend to deal in details. And so I think that does affect how you do science. I don't know why that is, it just seems that for me, and the other women scientists I've dealt with, we tend more to deal in the minute details, fine points."

Next to the theme of women's greater thoroughness, there was also the theme that women see the broader picture and do more *comprehensive work.* A successful woman scientist in our sample described her predisposition to produce complete and synthetic papers. Another woman observed, "Women tend to work longer on individual projects and take on projects that are broader in scope than do men. Women seem to find it more difficult to break projects into small parts, and consequently obtain fewer publications per project." In the words of a third woman scientist, "women tend to do more integrated work, maybe a little more big-picture stuff, and something that relates back to something that's relevant."

Publication Productivity Revisited

These findings suggest a reassessment of the often-observed publication productivity gap between the genders (for example, Cole and Zuckerman 1984; Fox 1983; Long 1992)—which we also found in our sample. The tendencies we have mentioned about thoroughness and comprehensiveness may combine to reduce women's quantitative publication output. If women scientists are more thorough and perfectionist than men, on average, and if they favor more comprehensive and synthetic work, the quantity of their publications per annum will tend to be lower.

Several women (but hardly any man) agreed that for women scientists, a higher quality of the individual paper counterbalanced the lower quantity of publications compared with men scientists. "Publications by women appear to be of higher quality with more data, more replications, more cautious interpretations and extension beyond data, and with greater depth into the question," commented one woman respondent. And another woman said, "I have observed men who do not do careful work and will report erroneous work. I have observed men who will publish, slightly rework a problem, and publish again. I have observed women who also publish nearly the same work in several articles. I personally do not publish work unless it is complete and I am very confident of the results."

One should emphasize that a perception, even if it is widespread, is not necessarily true—the low-quantity-but-high-quality claim may be a self-serving justification for low productivity. Nevertheless, some indirect quantitative evidence for the women scientists' tendency to publish articles that contain more substantial or comprehensive work emerged from a small study we made of the citations in the scientific literature to articles written by biologists (Sonnert 1991). Among the study's subsample of twenty-five former NSF fellows in biology who are now academic scientists, women's articles received significantly more citations per article, on average, than men's articles did (24.4 versus 14.4 citations; $p = 0.0337$). This greater impact might indicate that the women's articles tended to contain more noteworthy contents on the whole. A gender difference in citations per article in the same direction was also found by Long (1992) in a much larger sample of biochemists, as well as by Garfield (1993) in a study of the one thousand most-cited scientists.

This casts doubt on the appropriateness of the traditional indicator of a scientist's performance—publication productivity—which can be of great importance when decisions about scientists' careers are made. As a Project Access study among biologists confirmed, a long publication list appeared to be the most powerful determinant of quality judgments among peers, especially for a first impression (Sonnert 1991). If women as a group tend to have a slightly different publication behavior—less quantity but more quality—a performance measure based chiefly on publication counts may be biased against women.

Similar proportions of male and female interviewees did not believe that a high quantity of publications was an important part of being a good scientist (men: 54.7 percent; women: 59.4 percent; $p = 0.515$). This belief was only weakly related to the interviewees' actual publication output, but the direction of relationship differed between the genders. Among the women interviewees in academe, there was a weak positive correlation ($r = 0.16$) between believing that a high production output is important and the quantity of publications they produced. Among men, however, there was a small negative correlation ($r = -0.19$). The more publications the men interviewees produced, the less they tended to believe that a high publication productivity was an important part of being a good scientist. One might speculate that men, even if they rejected the quantitative performance standard, were nonetheless more prepared to abide by this standard for strategic reasons. Women, by contrast, might tend to be less willing to play the game of publication maximization if they considered it irrelevant for true quality—and this may contribute to the gender differential in average publication rates.

In another result on gender differences in the area of methodology, more men than women interviewees described their own overall scientific approach as creative (men: 26.9 percent; women: 11.8 percent; $p = 0.027$). The gender gap in the weight given to creativity is particularly intriguing because it runs counter to the common stereotype of the female being more creative and the male being more logical and analytical. Both our scientists' self-reports and their observations of others oppose this stereotype. A senior male scientist, for example, asserted that, owing to some deep-rooted gender differential, females were less creative and intuitive than males, which made them less suitable for high-quality work in his subspecialty. The responses in our group may, of course, reflect another stereotype about creativity—that men are the creative geniuses in all fields. Perhaps the comparatively strong emphasis on creativity among the men we interviewed reflects a powerful identification with the (male) heroes of science.

Standards of Good Science

Gender differences in the normative concept of good science may be an important element in bringing about different ways of doing science. We examined if our women interviewees as a group had a different idea, compared with their male counterparts, about what constitutes good science.

Addressing an important problem was seen as a major characteristic of really good scientific work by men and women interviewees equally (men: 34.8 percent; women: 33.0 percent). Among the interviewed former fellows, men emphasized creativity (men: 42.7 percent; women: 30.1 percent; $p = 0.070$) and good presentation of research (men: 11.2 percent; women: 3.9 percent; $p = 0.060$), whereas women stressed comprehensiveness (men: 20.2 percent; women: 35.9 percent; $p = 0.016$) and integrity (men: 4.5 percent; women: 13.6 percent; $p = 0.026$). In regard to the last aspect, a male scientist observed that women have "certainly kept their scientific integrity a lot more intact than some of their male counterparts. Few, if any, of the bozos involved in scientific fraud have been women, and hurrah for them, say I!"

As a demarcation criterion between top-quality and average work, women interviewees again emphasized comprehensiveness more than men did (men: 15.3 percent; women: 31.0 percent; $p = 0.011$). When asked about the characteristics of bad science, more women than men mentioned dishonesty (men: 9.3 percent; women: 23.8 percent; $p = 0.007$).

These findings did not indicate the existence of radical gender differences in what was considered good science; the women in our interview sample did not speak of a completely different set of quality criteria. But there were certainly gender differences in emphasis. These normative gender differences appeared to reflect the previously-described gender differences in perceived scientific styles and perhaps also in actual scientific behavior—as indicated in differential citation rates. For instance, the trend for women to do broader, more comprehensive work is echoed in a greater normative emphasis on comprehensiveness. The emphasis on integrity may be connected with women's tendency to be more thorough and perfectionist and to spurn "quick and dirty" work. It may also relate to women's less careerist approach to science or their unwillingness to cut corners for career success. These observations on the normative level bolster our comments on the causes of the productivity gap and of the higher citation rate of women's publications.

How are the normative views related to career outcomes? Our key finding, again, was the absence of strong correlations. Nonetheless, the following exceptions may be worth noting. Not surprisingly, successful academic scientists were more likely than less successful ones to emphasize creativity and originality as hallmarks of good science ($r = 0.18$; $p = 0.0766$); and successful women scientists, in particular, considered the lack of creativity a sign of bad science ($r = 0.41$; $p = 0.0048$). Moreover, successful scientists were more prone to define bad science as nonreplicable ($r = 0.20$; $p = 0.0621$)

Compared with the less successful women, outstanding women academics were marginally less interested in good presentation ($r = -0.25$; $p = 0.0856$) and instead emphasized advancement of knowledge ($r = 0.24$; $p = 0.0967$)—preferring content over form. Fewer successful men than undistinguished men mentioned throroughness and comprehensiveness ($r = -0.28$; $p = 0.0603$) as ingredients of excellence. By contrast, the noted tendency of women scientists to do thorough and comprehensive work might be particularly strong among the most eminent women. Among our women interviewees, characterizations of their own work as more thorough ($r = 0.33$; $p = 0.030$) and broader ($r = 0.34$; $p = 0.026$) than the mainstream correlated with success.

In sum, we found that the gender differences in scientific style, as reported by our respondents, lay far less in the epistemological and methodological than in the social aspects of science: for instance, collaboration with peers, competitiveness, problem selection, and professional conduct. The methodology was essentially the same for men and women, but women tended to be more meticulous and perfectionist in its use. Rather than being iconoclasts, women tended to uphold, to a particularly high degree, the traditional standards of science, such as carefulness, replicability, and connection to fundamentals. The women scientists in our sample seemed to have internalized the conventional epistemological rules of science more thoroughly than the social ones. In other words, women as strangers in science adopted—or were forced to adhere to—an extra-high measure of conformity to the traditional rules of science in terms of the formal way of conducting research. All the while, they were still standing somewhat on the margins in regard to the more informal aspects of social interactions and methods of career advancement.

Again, one should not forget that these gender differences are only trends; there is great variety in styles within each gender, and a large overlap between the genders.

We found little evidence that female practitioners of science followed or believed in a radically divergent epistemology or methodology that some feminist theorists of science have suggested. It may of course be proposed that women (and men) with alternative methodological and epistemological approaches do not flourish or survive in the science pipeline for very long, so that the scientists who are reasonably successful under the current system of science are predisposed to it, or at least have learned to accept it. But our sample of former postdoctoral fellows does not lend itself to testing such a proposition.

Women's tendency toward safe science in terms of seeking out niches as well as of methodological perfectionism may have various roots. From a difference-model perspective, females might be socialized to avoid conflict and to shy away from entering the fray with numerous competitors who all work on the same questions. They may take criticism more personally than men and therefore try harder to produce perfect work that is above any possibility of criticism. From a deficit-model perspective, the collegial environment may be particularly hostile to women who rock the boat. A woman scientist, for instance, reported that "there's always somebody watching for me to make a mistake." And another woman concurred that women scientists find themselves often "under the magnifying glass." A reaction to such extra scrutiny might be to tread gently and be extra careful by double- and triple-checking one's results and by using conventionally accepted methodology. Moreover, subtle or not so subtle exploitation of women's research work (such as downplaying their contribution or even appropriating their data) may have taught women scientists to protect their own research area. These two causes probably apply to different women scientists to different extents. What seems more important than gauging the relative weight of these explanations is to realize that they compound. Small individual and structural propensities combine to influence women scientists' problem choice and methodology.

The kick/reaction theory (Cole and Singer 1991) is useful in describing how structural and internal factors can work together in shaping individual career paths. Building on this theory, we should emphasize that different reactions are advantageous for different kinds of kicks: the appropriate strategy for countering negative kicks is resilience and hard work, gritting one's teeth in the face of obstacles and persisting. The appropriate strategy for taking advantage of positive kicks, by contrast, is being flexible in adjusting one's approach when new opportunities present themselves and taking risks on novel ideas.

As we have shown in our study, women scientists tended to receive a larger amount of negative kicks and a smaller amount of positive kicks. This may explain why women emphasized more than men the virtues of resilience and hard work, why women tended more toward safe science than men did, and why especially the successful women reported an extra-thorough working style. Owing to their career paths (in addition to potential prior dispositions), the women who survived in the science pipeline may have been conditioned to overcome negative kicks with appropriate responses—weathering them through resilience and hard work. They may have come to expect that, in an environment that institutionally has been less directed toward women's opportunities, they run a greater risk of being thwarted. Thus, the defensive approach of remaining in a safe niche appears more advantageous. Conversely, men's greater rate of exposure to positive kicks, on the average, may have taught them the value of following ideas that challenge conventional wisdom and may have given them the confidence to be more daring. This may eventually be an advantage for men in terms of their career advancement.

What has served a number of our women well in order to survive and become successful in science may also deter them from promoting ground-breaking scientific revolutions. Whereas some historians of science have pointed out that great scientific innovators were often marginal in relation to the mainstream science of the day (for a discussion, see Simonton 1988, 126–129), women's strangeness or marginality in science may have required and rewarded a tendency toward normal, or conventional, science. A double dose of marginality would put women in danger of being driven completely over the edge—out of science altogether. This is of course bound to change as women become less strange in the social system of science.

Science, Marriage, and Parenthood

So far we have primarily focused on science-internal elements of our respondents' career paths. Now we turn to a science-external area that is commonly considered to have a crucial impact on women's science careers in particular—marriage and parenthood. In family life, women are often faced with structural obstacles that men escape, and it is in family matters that distinctive cultural and socialization patterns internalized by women show themselves most clearly. In other words, both the deficit model and the difference model are needed to understand gender differences in science careers resulting from the experience of marriage and parenthood.

Most of the women interviewees have been married at one point in their lives, and the proportion of married men is even higher (men: 93.4 percent; women: 86.8 percent; $p = 0.118$). The majority of the women interviewees also have children, at about the same proportion as the men do (men: 71.4 percent; women: 70.2 percent). In contrast with earlier decades, when a larger proportion of women scientists were single, the problem of combining marriage and family with a career is a key issue facing most of our women interviewees.

We first checked in our questionnaire sample whether current marital and parental status (in various operationalizations) were related to the basic career outcomes (leaving science, employment area, academic rank, publication productivity). In general, we found that marital and parental statuses were unrelated to these career outcomes, both for men and—perhaps more surprising—for women. The minor exceptions were, first, that a high number of children weakly correlated with high academic rank for men and women ($p = 0.0520$; when gender was included as predictor). Second, a marginal interaction ($p = 0.0845$) in the expected direction indicated that married women respondents held a somewhat lower average rank than unmarried women did, whereas the opposite applied to the men.

If our overall analyses failed to show any strong interrelationships between the family and the career spheres for women scientists, does this mean there are none? We believe that interactions between family and science career do exist but that they have become too complex and idiosyncratic to be captured by global variables such as marital and parental statuses. Recall that we did find specific impacts of marriage and parenthood at the graduate and postdoctoral stages on later career outcomes. In the following, we survey how our respondents experienced the intersection of family and science career.

Among the small group of unmarried interviewees, a sizable minority of both women and men said that their decision not to marry was influenced by career demands (men: 27.3 percent; women: 34.6 percent; $p = 0.67$). Career considerations appeared to be more prevalent for married women than for married men in their decision not to have any children (men: 46.2 percent; women: 77.8 percent; $p = 0.152$).

A few male interviewees thought that men were more focused on their careers than women (men: 10.1 percent; women: 2.9 percent; $p = 0.050$); but when asked to assign percentages to the amount of priority given to science versus their private lives, men and women scientists attached similar weights and both ranked their careers higher in importance than their private lives (men: 61.0 percent; women: 64.4 percent). The group of women, however, may have maintained this career focus while facing some extra strain from the domestic sphere. Recall that a considerably larger percentage of the female questionnaire respondents mentioned family demands as a career obstacle (men: 2.8 percent; women: 21.3 percent; $p < 0.0005$). And whereas similar proportions of men and women interviewees had at some point contemplated abandoning science (men: 40.0 percent; women: 43.0 percent), the reasons were somewhat different by gender. Slightly more women than men mentioned familial and marital considerations (men: 7.3 percent; women: 17.3 percent; $p = 0.14$), while the lack of recognition, success, or money was more prevalent among men than women (men: 31.7 percent; women: 17.3 percent; $p = 0.11$).

When those who were married were asked if their marriage had affected their careers, almost everyone said yes (men: 91.4 percent; women: 91.8 percent). Contrary to our intuitive sense of marriage as a career obstacle for women, many women reported career advantages that derived from being married. The effect of marriage on their career was considered positive by almost half the married respondents, both men and women (men: 45.7 percent; women: 49.4 percent; $p = 0.633$). Only a small group mentioned an explicitly negative impact (men: 14.8 percent; women: 17.6 percent; $p = 0.624$). Both men and women regarded the emotional support and security of marriage as one of its key advantages. A woman, for instance, noted that marriage "made me more secure, it let me devote myself to my work in ways I might not otherwise have, because my attention would have been diverted to finding a mate, or finding a companion, and being married helped me be settled."

A likely scenario, especially for women scientists, is to be married to another scientist, often in the same field (Fava & Deierlein 1988). In our questionnaire sample of 699 former postdoctoral fellows, 62.0 percent of the married women but only 18.9 percent of the married men had a spouse with a doctorate. Spouses who are also scientists were often described as understanding and supportive of the time-consuming and work-dedicated life-style of scientists. A male scientist recounted how his first marriage floundered largely because of his rigorous work schedule; in his second marriage, to a scientist in his field, "it's o.k. if I work ten hours a day. I'm married to someone now who also likes to work ten hours a day, so it's not a problem." A spouse in the same field is also a sounding board for ideas, a source of inspiration and criticism. Spouses in the same field often collaborate, but the potential pitfall for the woman is that she might be perceived as her husband's assistant and junior partner in their collaborative efforts.

The negative aspects of marriage include, first and foremost, restrictions in mobility. Because women scientists are much more likely than men to live in two-scientist marriages, the problem is more prevalent among women scientists. More women than men interviewees thought that geographical issues were a factor in their careers (men: 62.6 percent; women: 75.5 percent; $p = 0.0534$). In addition, the geographical issues were typically solved at the wife's expense. Many women compromised their own career by following a husband to his job location or by being stuck in a place unfavorable for their own career because the husband was not geographically mobile. A few of our interviewees resorted to some form of long-distance marriage. In an extreme case, a woman interviewee and her husband, a scientist in the same field, lived on different continents and saw each other mostly at conferences.

The interviewees who were single noted the obvious advantages of their status—more time to work and the chance to make optimal career choices unrestricted by spouse or family considerations. On the negative side, there was the lack of emotional and even practical support in managing the chores of everyday life. As a single woman respondent put it, "it is very clear, as my other single friends and I agree, everybody needs a spouse, somebody to do things for them, if only to take the cat to the vet."

While those corollaries of unmarried life apply to both men and women, single women also face a set of unique disadvantages within the social system of science. A single female scientist can have a particularly awkward standing in a predominantly male environment (Kaufman 1978). She might be considered "available" and thus be more likely to attract unwanted sexual attention from colleagues as well as hostility from colleagues' wives. A male respondent observed "enormous pressure on an unattached woman scientist to date her colleagues, and no pressure for a comparable male scientist." A single woman scientist might also deter male colleagues from socially interacting with her for fear of a misinterpretation of their relationship. A divorced woman interviewee noted that by "having a man anywhere in the picture instead of always standing on my own, I would have had substantially more clout. Because then at those departmental parties or whatever, when the scientists were bringing their wives, I would have brought my husband." Finally, single women scientists might be viewed as wives- and mothers-to-be and thus be unable to escape any potential prejudice against married women scientists.

Gender differences are substantial when it comes to the perceived effects of parenthood. Almost all the women interviewees with children said that being a parent had somehow influenced their careers, but only two-thirds of the men felt this way (men: 65.2 percent; women: 93.1 percent; $p = 0.0001$). Whereas some of the interviewees with children noted a positive influence of children on their careers (men: 17.5 percent; women: 19.0 percent), a slightly higher percentage noted negative influences (men: 25.0 percent; women: 31.7 percent); and the largest group stated conflicting or unclear effects (men: 55.0 percent; women: 47.6 percent).

As a major negative career effect of parenthood, the interviewees noted that time and energy is taken away from scientific pursuits. A number of women noted that the child-rearing years tend to coincide with the important early career phase of establishing scientific credentials (for instance, during the tenure-track period). Negative career effects of parenthood were reported not only by women but also by men. A male scientist, for example, who found himself in the somewhat uncommon role of being a single father of four young children, noted that his scientific career went into a dip during that time. Another male interviewee said he was "starting to turn things down because I hate to see my older boy disappointed when I'm not there to tuck him in." On the other hand, interviewees also said that children provided an intense emotional satisfaction that put their career in perspective and even helped them to function as a scientist, as this woman pungently professed: "I think that if I had to look at myself in terms of my accomplishments entirely in terms of my career, I would be a sad cookie. And that these setbacks and stupid remarks, and the crap that gets thrown at you, would bother me a s——tload more if I had also sacrificed the experience of having a child."

The disadvantages of marriage and parenthood for women's careers appear straightforward and intuitively plausible—restricted mobility for dual-career couples, for example, or the large amounts of time and effort required by child rearing. The advantages, by contrast, seem to be more indirect and subtle; it is hard to gauge the impact of the social and emotional support that a spouse provides. The greater visibility of the disadvantages may be a source for the widespread view of marriage and family as career obstacles for women—and, indeed, a substantial fraction of our female respondents named these areas as obstacles in their careers. But when we asked our interviewees specifically about the effects of marriage and parenthood, a much more varied and complex picture of the interrelation between domestic sphere and career emerged. This may help explain why there was hardly any correlation between career outcomes and marital and parental statuses for women (as well as for men).

It seems impossible to summarize our results into one simple statement about the effect of marriage and parenthood. First, the style of marriage is important. If, in a traditional mode of marriage, the wife has to take care of the domestic chores, the husband is an extra burden. A woman interviewee who divorced her husband and raised her child by herself commented, "When I left my husband, it was as if I had two children less." A more cooperative mode of marriage, by contrast, may ease the wife's burdens and provide her with additional support. Moreover, the profession and seniority of the spouse are crucial factors. Even in our group of interviewees comprising many distinguished women scientists, the majority of married women had husbands who were the primary wage earners. Of the married women, only 34.2 percent reported to be the primary wage earner in the family, but 80.3 percent of the married men did ($p < 0.0001$). A general marriage pattern in our culture is for women to marry men who are older and professionally senior and who earn more money than themselves, which encourages the couple to give the husband's career priority. It is not surprising that an advantageous marriage choice for women might be a nontraditional pattern of marrying a junior or equal-status scientist or a "trailing spouse" (CSWP 1992a, 12) whose job makes him geographically mobile. Being the main breadwinner has financial but also psychological effects. As a woman scientist said, being the primary wage earner "helped reinforce my being determined" in pursuing career success.

Rather than thinking of marriage and parenthood as having a fixed effect on women scientists' careers, we should see marriage and parenthood as a set of problems and opportunities. Women scientists are faced with the dilemma of "synchronizing" the often conflicting demands of three clocks: their biological clock, their career clock (such as their tenure clock), and their spouses' career clock. On the other hand, a husband and a family can provide emotional security and financial stability, as well as scientific support if the husband is a scientist in the same field. Largely depending on how the problems are resolved and the opportunities are used, the effect of marriage and parenthood on women scientists' careers may be positive or negative. In other words, marital and parental statuses did not appear as structural forces that inescapably pushed our respondents in one direction. Rather, our respondents could make choices about how to deal with the constraints *and* opportunities. Some choices turned out to be more fortuitous than others.

There are women among our interviewees who combined a family with a successful career, but there are also women who had both a troubled career and difficulties in private life. One woman, for instance, made several career sacrifices to accommodate her husband, who eventually terminated their marriage anyway, and she was left in an insecure job. The constraints of family obligations and the availability of a wide scope of options (ranging from family to career orientation) are not mutually exclusive explanations for women's choices in balancing family and career. Rather, they often seem to combine and reinforce each other. In some cases, family obligations created difficulties for a woman's career that then made her opt to focus on her family. When a woman respondent's marriage, for instance, limited her career chances, she tried to make the best of it by scaling back the priority of her science career and starting a family while working part time. Other women we interviewed also responded to difficulties in one area by concentrating on the other. For some, disappointment in a career led to a greater focus on the family, but this mechanism also works the other way around. A never-married woman interviewee described how she became a career woman "by default." If she had met a suitable husband, she indicated she would have willingly reduced her career aspirations.

A man is still widely expected to have a successful career and be the breadwinner of his family so that going home to be a house husband is an unconventional and rare option. Nonetheless, some among our male interviewees expressed a desire, if not to become house husbands, at least to be more involved with their families. One male interviewee, for instance, emphasized that he quit his academic position to have more time for his family. The trend widening the range of women's options in terms of family life or career has facilitated women's participation in all kinds of professional careers. This trend may eventually also widen the range of options for men and allow them to focus more strongly than at present on their family life. In other words, if the social system of science becomes more sensitive toward family responsibilities—in terms of parental leave, part-time options, and so on—this will benefit not only women but also men who want to get more involved in their families.

5

CONCLUSION

IN CONCLUSION, LET US return to Simmel's concept of the stranger. What do our results contribute to answering the central question—to what extent are women still strangers in science? After reviewing our results, we will put the situation encountered in science into its wider societal framework.

Overall, the condition of women in the sciences has improved considerably during the past few decades, and several women scientists in the Project Access study explicitly attested to sweeping changes in the position of women in science. Although this study has focused on differences between men and women scientists as groups, we reemphasize that there are great variations in individual career paths and obstacles within each gender. The presented gender differences and gender-related obstacles in science are tendencies rather than distinct dividing lines that put every woman scientist on one side and every man scientist on the other.

A major motivation for this study was to find out whether the career paths of women scientists followed a glass-ceiling or a threshold pattern. Studying a select subgroup of American scientists—former recipients of prestigious postdoctoral fellowships—allowed us to pose the question of glass ceiling versus threshold more precisely than is usually possible, comparing men and women scientists with similar auspicious circumstances at the beginnings of their professional careers. We reiterate that this group of scientists is not representative of all American doctorate-level scientists so that our results do not necessarily apply to scientists in general.

On the whole, the women in our sample did not do extremely worse than the men; very large and very obvious gender differences and disparities were absent. While the gender gap has narrowed, however, full gender equality in science careers and women's full "ownership," alongside men, of science still seemed elusive. A highly complex mosaic of gender relations and gender differences has replaced the once clear-cut gender division of labor in science.

A key result was that field mattered. Whereas the careers of women biologists clearly followed a threshold pattern, a glass ceiling became obvious in the other sciences. This suggested to us a critical mass effect. To some degree, women's social marginalization may be a function of sheer numbers. Because of their higher numbers, women biologists might not be considered outsiders or strangers to the extent that their female colleagues in other sciences are.

To explain the glass ceiling in more detail, we employed the theoretical notions of the deficit model and the difference model. The deficit model focused on structural forces—women are *treated* as strangers in science—whereas the difference model focused on women's internal differences—women *act* as strangers in science.

In terms of the deficit model, many women experienced discrimination at various levels of seriousness. Because outright discrimination may subside, however, attention needs to be shifted to more elusive mechanisms. Subtle obstacles, such as a marginalization in the network of collegial interaction, were frequently reported. Furthermore, factors typical of the difference-model may become more prominent—letting some gender disparities persist even in the absence of structural handicaps. In terms of the difference model, we found differences in self-confidence, ambition, career goals, and interaction style—each slight but in the same direction. And a somewhat different outlook on the normative concept of good science accompanied women's slightly different style of doing science. Structural impacts (of the deficit model) and internal predispositions (of the difference model) are not isolated but very likely combine to bring about differences in career achievement between the genders.

A focus on earlier career phases allowed us to determine gender-specific differences in the accumulation of human capital—of advantages and disadvantages. A number of such differences were found, but they did not all indicate a compounding of disadvantage for our group of women. In some cases, women as a group also made choices that anticipated and circumvented disadvantages. Findings about the impact of marriage and family on science careers reemphasized this theme. Whereas marital and parental status per se were little connected to career outcomes, a number of specific corollaries did make a difference, such as taking the postdoctoral fellowship in order to be with one's spouse or becoming a parent during the postdoctoral fellowship. Therefore, individual scientists, women as well as men, should not be viewed as buffeted by structural forces. They seem to have a great amount of leeway in making choices—in handling the problems as well as the opportunities of these statuses. Methodologically, this means for social science that global correlations between characteristics and outcomes may become even weaker than is customary; and researchers may have to pay increasing attention to the fine structure of individuals' experiences and the dynamics of their life stories.

What we have traced in this study—women as strangers making their way into science—is just one instance of a pervasive phenomenon in modern society. Women's experience in science may parallel their experience in other fields. Moreover, other types of strangers—racial, ethnic, or religious minorities, for instance, or those from underprivileged classes—may also have similar experiences when entering areas previously closed to them. With the waning of clear-cut ascriptive statuses, strangers become more alike; but everyone is exposed to strangers much more than before.

In a society as open and diverse as the United States, social institutions will face an increasing influx of strangers of many kinds—some from fading old social divisions, some from newly developing ones. Although there is a general consensus that structural obstacles typical of the deficit model should be minimized, the most intriguing issues, for research as well as for policy-making, lie in the area of the difference model. Will strangers be eventually assimilated, thereby losing their differences; or will they be integrated, thereby causing some change in the host institutions? In other words, will strangers be considered liabilities or assets?

APPENDIX

The Questionnaire Sample

The key variables of gender, year of fellowship, and academic field were known for every recipient of an NSF postdoctoral fellowship. To form a general impression of the representativeness of our sample, we used these variables to compare the active population (all those who could have been part of our study, according to our information) with the contacted sample (those who could be located and were sent questionnaires) and with our actual respondents.

In respect to gender, the three groups were very similar, although the proportion of men slightly increased from the active population to the contacted sample and to the respondents (table A.1).

The distributions of the year of fellowship, again, are very similar for the active population, the contacted sample, and the respondents (table A.2).

There were slightly fewer biologists and slightly more mathematicians and physical scientists in the contacted sample, compared with the active population (table A.3). The share of biologists continued its minor decline in the step from the contacted sample to the respondents. Conversely, the percentage of physical scientists increased again in this step. These differences, however, were far from dramatic so that we can consider the distributions similar.

Examining the field distribution by gender (table A.4), we found that the men were primarily responsible for the noted fluctuations in both the biological and the physical sciences, whereas the proportion of women remained virtually unchanged across the three groups. The percentage of women in computer science and mathematics decreased

Table A.1. Questionnaire Sample by Gender

	Active population	Contacted sample	Respondents
Men	887 (75.3%)	581 (76.3%)	361 (78.3%)
Women	291 (24.7%)	180 (23.7%)	100 (21.7%)
N	1178 (100.0%)	761 (100.0%)	461 (100.0%)

Table A.2. Questionnaire Sample by Year of Postdoctoral Fellowship

	Active population	Contacted sample	Respondents
1955–1974	324 (27.5%)	232 (30.5%)	126 (27.3%)
1975–1979	517 (43.9%)	316 (41.5%)	199 (43.2%)
1980–1984	270 (22.9%)	170 (22.3%)	110 (23.9%)
1985+	67 (5.7%)	43 (5.7%)	26 (5.6%)
N	1178 (100.0%)	761 (100.0%)	461 (100.0%)

Table A.3. Questionnaire Sample by Field of Fellowship

	Active population	Contacted sample	Respondents
Agriculture	5 (0.4%)	3 (0.4%)	2 (0.4%)
Biological sciences	515 (43.7%)	312 (41.0%)	181 (39.3%)
Health Sciences	5 (0.4%)	4 (0.5%)	3 (0.7%)
Engineering	31 (2.6%)	23 (3.0%)	15 (3.3%)
Computer science and math	194 (16.5%)	138 (18.1%)	83 (18.0%)
Physical sciences	234 (19.9%)	164 (21.6%)	110 (23.9%)
Social sciences	170 (14.4%)	101 (13.3%)	58 (12.3%)
Humanities	24 (2.0%)	16 (2.1%)	9 (2.0%)
N	1178 (99.9%)	761 (100.0%)	461 (99.9%)

Table A.4. Questionnaire Sample by Field of Fellowship and Gender

	Active population		Contacted sample		Respondents	
	Men	*Women*	*Men*	*Women*	*Men*	*Women*
Agriculture	0.6%	0.0%	0.5%	0.0%	0.5%	0.0%
Biological sciences	39.5%	56.7%	37.0%	55.0%	34.9%	55.0%
Health sciences	0.3%	0.7%	0.3%	1.1%	0.5%	1.0%
Engineering	3.4%	0.3%	3.8%	0.6%	3.9%	1.0%
Computer science and math	19.3%	7.9%	21.2%	7.8%	21.6%	5.0%
Physical sciences	22.5%	11.7%	24.1%	12.8%	26.9%	13.0%
Social sciences	13.1%	18.6%	11.9%	17.8%	10.2%	21.0%
Humanities	1.4%	4.1%	1.2%	5.0%	1.4%	4.0%
	100.1%	100.0%	100.0%	100.1%	99.9%	100.0%

somewhat from the contacted sample to the respondents, whereas the percentage of their male counterparts increased from the active population to the contacted sample. Female social scientists were slightly over-represented among the respondents relative to the contacted sample, and the proportion of male social scientists steadily decreased. Nevertheless, all these deviations were minor.

We concluded that the inspection of the three key variables did not reveal any major bias among our respondents as compared with the contacted sample or the active population.

NOTES

Chapter 1

1. A number of scholars believe that a genetic component underlies gender differences (such as Rossi 1977). For example, some claim that a gender-related genetic factor gives males superior mathematical abilities (Benbow and Stanley 1980; Hoben 1985). This view has been vigorously criticized by those who see early socialization as the cause of the gender differential in mathematical ability (Campbell and Geller 1984; Eccles and Jacobs, 1986; Entwisle and Hayduk 1988). In their landmark survey of studies of gender differences, Maccoby and Jacklin (1974) concluded that convincing evidence for a genetically determined gender difference existed in only two areas: the average male is more aggressive and has a better visual-spatial ability than the average female. By contrast, female superiority in measures of verbal ability and male superiority in measures of mathematical ability do not appear to have a genetic basis.

2. To control for cohort and field effects, we included academic age and fields in the regression analyses of our questionnaire sample. For our interview sample, gender differences are reported "straight" because this sample is to a high degree, albeit not perfectly, matched for academic age and fields.

3. Compared with the former postdoctoral scientists in other fields, our group of social scientists was the smallest and the least typical of doctoral scientists in their field. Therefore, our analyses did not particularly focus on the social scientists.

4. Before 1975, men outnumbered women among the NSF postdoctoral fellows roughly 20:1. Therefore, we modified the selection procedure for this group. We attempted to contact all women who were fellows before 1975 in the usual manner. A contacted sample of sixty-nine women yielded thirty-six respondents (response rate of 52.2 percent). Then we matched the thirty-six women respondents with the men awardees of the same academic field and year of fellowship and attempted to contact only these men. A contacted sample of 163 men yielded ninety respondents (response rate of 55.2 percent).

Chapter 2

1. Academic rank is subject to a ceiling effect. The great majority of scientists who spend their entire professional lives in academe can expect to reach the position of full professor eventually, and scientists who accomplish this goal early in their careers cannot advance further. Thus, academic position is a better indicator of intraprofessional differentiation in the first couple of decades of an academic career than it is later on. Most of the respondents in our sample are younger or middle-aged scientists for whom academic rank is a meaningful indicator of academic success.

2. Some positions in academe, such as department chair and dean, could arguably be located above full professorships. Most scientists, however, appear to consider full professor to be the top rank in a regular academic career. The other positions are regarded as combinations of scientific and administrative jobs.

3. It would have been interesting to check whether former NRC associates were more successful than former NSF fellows in *non*academic science positions, but we had no way of assessing success in nonacademic science.

4. We found a correlation of 0.714 between the raw publication rate and a more refined measure based on Sabljic and Trinajstic's (1988) formula, which takes into account the journals' impact rating as well as the number of authors in collaborative publications. Journal impact ratings, published by the Institute for Scientific Information in Philadelphia, are calculated from the citations to a journal's articles.

5. Attempts to show gender disparities in multiple regression may be vulnerable to criticism by what can be called the reverse-regression argument. Dempster (1984, 1988) illustrated that, under certain circumstances, radically different interpretations can be drawn from the same data set depending on the choice of the dependent variable. In one of Dempster's examples, women were found to earn less money than men did at each experience level—which could be taken as an indication of gender discrimination. But after reversing the regression, the same data showed that, at each level of earnings, women had less experience than men did—which could be taken as evidence of reverse discrimination. In our situation, rank and publication productivity are in a similar relationship. If women are at a lower rank for each level of productivity, this may indicate that women are disadvantaged; but this finding might be challenged if it was also true that, at each rank, women were less productive. In our data, we found a nonsignificant gender difference in productivity, controlling for rank, but a significant gender difference in rank, controlling for productivity. Thus, our finding that women are disadvantaged in rank, even after controlling for publication productivity, escapes the possible challenge of the reverse-regression argument.

Chapter 3

1. 1 = less than a high school diploma; 2 = high school diploma; 3 = some undergraduate college; 4 = associate's degree (A.A.); 5 = bachelor's degree (B.S., B.A., A.B., B.B.A., B.S.E.); 6 = some graduate work but no advanced degree; 7 = master's degree (M.S., M.Sc., M.A., S.M., M.Ed., M.F.A.) or professional postgraduate degree (J.D., L.L.B., M.B.A., D.Pharm.); 8 = post-master's work but no doctorate; 9 = doctoral degree (Ph.D., D.Sc., Ed.D., M.D., D.D.S.).

2. The expected number of firstborns, given sibship size, is calculated as follows: (number of people in two-sibship families × probability of being firstborn in two-sibship families) + (number of people in three-sibship families × probability of being firstborn in three-sibship families) and so on. In our case: $198/2 + 184/3 + 108/4 + 50/5 + 23/6 + 11/7 + 8/8 + 7/9 + 2/10 + 4/11 = 205.08$.

3. Similar results were obtained when examining the gender differences in the incidence of parental death or divorce by age twelve instead of age eighteen.

4. Part-time students are defined as those who spent at least one year between bachelor and doctorate as part-time students. Interrupters spent at least one year between bachelor and doctorate not working on their degrees. Thus, interruption covers both a pause during graduate studies and a delay between attaining the bachelor's degree and commencing graduate studies.

5. For each significant first-principal component, the stepwise-regression approach usually yielded one significant variable among the simple variables forming the component. Some of the gender interactions identified as significant by stepwise regressions lost their significance in the final model. They were reported nonetheless.

6. 1 = negative; 2 = no influence (neutral); 3 = slightly positive; 4 = moderately positive; 5 = very positive.

Chapter 4

1. The fashionable word *serendipity* has a somewhat complex etymology (Goodman 1961). In one usage, it describes the happy coincidence when a person who is actively looking for one thing finds another. Our respondents, however, commonly used serendipity in a slightly different sense—just being in the right place at the right time.

2. Thus, a higher percentage of women interviewees than of women questionnaire respondents noted discrimination as a career obstacle. As we have already mentioned, the percentage for the women questionnaire respondents was 12 percent. This may have to do with the different composition of the two samples. Possibly the different format—a conversation with a female interviewer—increased the interviewees' propensity to share such an experience.

3. Alphabetical author listings do, of course, contaminate this indicator. In these cases, the first author might be the main contributor—or just the one whose name comes first in the alphabet. We assume that alphabetical listings occur equally for women's and men's publications so that, in our large sample, a comparison by gender of the proportion of first-authorships is still meaningful.

4. In physics, it was Enrico Fermi who started to apply the term *family* to teams (Holton 1978).

REFERENCES

Adams, B. N. (1972). Birth order. A critical review. *Sociometry, 35,* 411–439.

Allison, P. D., & Long, J. S. (1987). Interuniversity mobility of academic scientists. *American Sociological Review, 52,* 643–652.

Allison, P. D., & Long, J. S. (1990). Departmental effects on scientific productivity. *American Sociological Review, 55,* 469–478.

Allison, P. D., & Stewart, J. A. (1974). Productivity differences among scientists: Evidence for accumulative advantage. *American Sociological Review, 39,* 596–606.

Belenky, M. F., Clinchy, B. M., Goldberger, N. R., & Tarule, J. M. (1986). *Women's ways of knowing: The development of self, voice, and mind.* New York: Basic Books.

Benbow, C. P., & Stanley, J. C. (1980). Sex differences in mathematical ability: Fact or artifact. *Science, 210,* 1262–1264.

Bernard, J., (1964). *Academic women.* University Park: Pennsylvania State University Press.

Bielby, W. T., & Baron, J. N. (1986). Men and women at work: Sex segregation and statistical discrimination. *American Journal of Sociology, 91,* 759–799.

Blau, F. D., & Jusenius, C. L. (1976). Economists' approaches to sex segregation in the labor market: An appraisal. *Signs, 1,* 181–199.

Brim, O. G. (1958). Family structure and sex role learning by children. *Sociometry, 21,* 1–16.

Briscoe, A. M. (1984). Scientific sexism: The world of chemistry. In V. B. Haas & C. C. Perrucci. (Eds.), *Women in scientific and engineering professions* (pp. 147–159). Ann Arbor: University of Michigan Press.

Brush, S. G. (1991). Women in science and engineering. *American Scientist, 79,* 404–419.

Campbell, P. F., & Geller, S. C. (1984). Early socialization: Causes and cures of mathematics anxiety. In V. B. Haas & C. C. Perrucci (Eds.), *Women in scientific and engineering professions* (pp. 173–180). Ann Arbor: University of Michigan Press.

Centra, J. A. (1974). *Women, men and the doctorate.* Princeton, NJ: Educational Testing Service.

Chamberlain, M. K. (Ed.). (1988). *Women in academe: Progress and prospects.* New York: Russell Sage.

Chodorow, N. (1974). Family structure and feminine personality. In M. Z. Rosaldo & L. Lamphere (Eds.), *Women, culture and society* (pp. 43–66). Stanford: Stanford University Press.

Clark, R. D., & Rice, G. A. (1982). Family constellation and eminence: The birth orders of Nobel Prize winners. *Journal of Psychology, 110*, 281–287.

Clemente, F. (1973). Early career determinants of research productivity. *American Journal of Sociology, 79*, 409–419.

Cole, J. R. (1979). *Fair science: Women in the scientific community.* New York: Free Press.

Cole, J. R. (1987). Women in science. In D. N. Jackson & J. P. Rushton (Eds.). *Scientific excellence. Origins and assessment.* (pp. 359–375). Newbury Park, CA: Sage.

Cole, J. R., & Singer, B. (1991). A theory of limited differences: Explaining the productivity puzzle in science. In H. Zuckerman, J. R. Cole, & J. T. Bruer. (Eds.), *The outer circle: Women in the scientific community* (pp. 277–310). New York: Norton.

Cole, J. R., & Zuckerman, H. (1984). The productivity puzzle: Persistence and change in patterns of publication of men and women scientists. In M. W. Steinkamp & M. L. Maehr (Eds.), *Advances in motivation and achievement* (Vol. 2), (pp. 217–256). Greenwich, CT: JAI Press.

Cole, S., & Fiorentine, R. (1991). Discrimination against women in science: The confusion of outcome and process. In H. Zuckerman, J. R. Cole, & J. T. Bruer. (Eds.), *The outer circle: Women in the scientific community* (pp. 205–226). New York: Norton.

CSWP (1992a). CSWP grapples with climate, dual career issues. *APS News, 1*(7), 9–15.

CSWP (1992b). Twenty years later, CSWP is still going strong. *APS News, 1*(6), 11–16.

Dempster, A. P. (1984). Alternative models for inferring employment discrimination from statistical data. In P.S.R.S. Rao & J. Sedransk (Eds.), *W. G. Cochran's impact on statistics* (pp. 309–330). New York: Wiley.

Dempster, A. P. (1988). Employment discrimination and statistical science. *Statistical Science, 3*(2), 149–161.

Dresselhaus, M. S. (1986, June). Women graduate students. *Physics Today, 39*, 74–75.

Eccles, J. S. (1987). Gender roles and women's achievement-related decisions. *Psychology of Women Quarterly, 11*, 135–172.

Eccles, J. S., & Jacobs, J. E. (1986). Social forces shape math attitudes and performance. *Signs, 11*, 367–380.

Eccles-Parsons, J., Adler, T. F., & Kaczala, C. M. (1982). Socialization of achievement attitudes and beliefs: Parental influences. *Child Development, 53*, 310–321.

Eisenstein, H. (1983). *Contemporary feminist thought.* Boston: G. K. Hall.

Entwisle, D. R., & Hayduk, L. A. (1988). Lasting effects of elementary school. *Sociology of Education, 6*, 147–159.

Epstein, C. F. (1970). *Woman's place: Options and limits in professional careers.* Berkeley: University of California Press.

Fava, S. F., & Deierlein, K. (1988, August). Women physicists in the U.S. The career influence of marital status. *Gazette: A Newsletter of the Committee on the Status of Women in Physics of the American Physical Society, 8*(2), 1–3.

Ferber, M., & Huber, J. (1979). Husbands, wives, and careers. *Journal of Marriage and the Family, 41*, 315–325.

Ferber, M. A., & Loeb, J. W. (1973). Performance, rewards, and perceptions of sex discrimination among male and female faculty. *American Journal of Sociology, 78*, 995–1002.

Fiss, O. M. (1991). An uncertain inheritance. In H. Zuckerman, J. R. Cole, & J. T. Bruer (Eds.), *The outer circle: Women in the scientific community* (pp. 259–273). New York: Norton.

Folger, J. K., Astin, H. S., & Bayer, A. E. (1970). *Human resources and higher education: Staff report of the commission on human resources and advanced education.* New York: Russell Sage.

Fox, M. F. (1983). Publication productivity among scientists: A critical review. *Social Studies of Science, 13*, 285–305.

Fox, M. F. (1991). Gender, environmental milieu, and productivity. In H. Zuckerman, J. R. Cole, & J. T. Bruer (Eds.), *The outer circle: Women in the scientific community* (pp. 188–204). New York: Norton.

Fox, M. F., & Ferri, V. C. (1992). Women, men, and their attributions for success in academe. *Social Psychology Quarterly, 55*, 257–271.

Frieze, I. H. (1978). Psychological barriers for women in sciences: Internal and external. In J. A. Ramaley (Ed.), *Covert discrimination and women in the sciences* (pp. 65–95). Boulder, CO: Westview.

Frieze, I. H., Fisher, J., Hanusa, B., McHugh, M. C., & Valle, V. A. (1978). Attributions of the causes of success and failure as internal and external barriers to achievement in women. In J. A. Sherman & F. L. Denmark (Eds.), *Psychology of women: Future directions in research* (pp. 519–552). New York: Psychological Dimensions.

Galton, F. (1875). *English men of science: Their nature and nurture.* New York: Appleton.

Garfield, E. (1993, March 1). Women in science. Part 1: The productivity puzzle—J. Scott Long on why women biochemists publish less than men. *Current Comments, 9*, 3–5.

Gilbert, G. N., & Mulkay, M. (1984). *Opening Pandora's box: A sociological analysis of scientists' discourse.* Cambridge, UK: Cambridge University Press.

Goodman, L. A. (1961). Notes on the etymology of *serendipity* and some related philological observations. *Modern Language Notes, 76*, 454–457.

Hagstrom, W. O. (1971). Inputs, outputs and the prestige of university science departments. *Sociology of Education, 44*, 375–397.

Hall, R. M. (1982). *The classroom climate: A chilly one for women?* (Project on the status and education of women). Washington, DC: Association of American Colleges.

Heikkinen, H. (1978). Sex bias in chemistry texts: Where is women's place? *The Science Teacher, 45*, 16–21.

Hoben, T. (1985). A theory of high mathematical aptitude. *Journal of Mathematical Psychology, 29*, 231–242.

Hochschild, A. R. (1989). *The second shift: Working parents and the revolution at home.* New York: Viking.

Hoffnung, M. (1984). Motherhood: Contemporary conflict for women. In J. Freeman (Ed.), *Women: A feminist perspective* (3rd ed.), (pp. 124–138). Palo Alto, CA: Mayfield.

Holton, G. (1978). On the psychology of scientists, and their social concerns. In *The scientific imagination: Case studies* (pp. 229–254). Cambridge, UK: Cambridge University Press.

Hornig, L. S. (1984). Professional women in transition. In V. B. Haas & C. C. Perrucci (Eds.), *Women in scientific and engineering professions* (pp. 43–58). Ann Arbor: University of Michigan Press.

Hornig, L. S. (1987). Women graduate students. In L. S. Dix (Ed.), *Women: Their underrepresentation and career differentials in science and engineering* (pp. 103–122). Washington, DC: National Academy Press.

Huston-Stein, A., & Higgins-Trenk, A. (1978). Development of females from childhood through adulthood: Careers and feminine role orientations. *Life-span Development and Behavior, 1*, 257–296.

Jasanoff, S. (1992). Pluralism and convergence in international science policy. In N. Keyfitz (Ed.), *Science and sustainability* (pp. 157–180). Laxenburg, Austria: International Institute for Applied Systems Analysis.

Jones, L. V., Lindzey, G., & Coggeshall, P. E. (Eds.). (1982). *An assessment of research-doctorate programs in the United States* (5 vols.). Washington, DC: National Academy Press.

Kagan, J. (1971). *Personality development.* New York: Harcourt Brace Jovanovich.

Kanter, R. M. (1977a). *Men and women of the corporation.* New York: Basic Books.

Kanter, R. M. (1977b). Some effects of proportions on group life: Skewed sex ratios and responses to token women. *American Journal of Sociology, 82*, 965–990.

Kaufman, D. R. (1978). Associational ties in academe: Some male and female differences. *Sex Roles, 4*, 9–21.

Keller, E. F. (1983). *A feeling for the organism: The life and work of Barbara McClintock.* New York: W. H. Freeman.

Keller, E. F. (1985). *Reflections on gender and science.* New Haven: Yale University Press.

Keller, E. F. (1989). Feminism and science. In A. Garry & M. Pearsall (Eds.), *Women, knowledge, and reality: Explorations in feminist philosophy* (pp. 175–188). Boston: Unwin Hyman.

Kelly, A. (1985). The construction of masculine science. *British Journal of Sociology of Education, 6*, 133–154.

Keohane, N. (1984, January 25). *Women and power.* Talk at the Cambridge Forum. Audiotape No. 544.

Kerr, P. (1988). *A conceptualization of learning, teaching and research experiences of women scientists and its implications for science.* Unpublished Ph.D. dissertation, Cornell University, Ithaca NY.

Kistiakowsky, V. (1980, February). Women in physics: Unnecessary, injurious and out of place? *Physics Today, 33*, 32–40.

Kjerulff, K. H., & Blood, M. R. (1973). A comparison of communication patterns in male and female graduate students. *Journal of Higher Education, 44*, 623–632.

Knorr-Cetina, K. D. (1981). *The manufacture of knowledge: An essay on the constructivist and contextual nature of science.* Oxford: Pergamon.

Koshland, D. E. (1988). Women in science. *Science, 239*, 1473.

Kreckel, R. (1992). The concept of class: Its uses and limitations in the analysis of social inequality in advanced capitalist societies. In S. Kozyr-Kowalski & A. Przestalski (Eds.). *On social differentiation: A contribution to the critique of marxist ideology* (Part 1, pp. 29–51). Poznan, Poland: Adam Mickiewics University Press.

Kreckel, R. (1993). Doppelte Vergesellschaftung und geschlechtsspezifische Arbeitsmarktstrukturierung. In P. Frerichs & M. Steinrücke (Eds.). *Soziale Ungleichheit und Geschlechterverhältnisse* (pp. 51–63). Opladen, Germany: Leske & Budrich.

Latour, B., & Woolgar, S. (1979). *Laboratory life: The social construction of scientific facts.* Beverly Hills, CA: Sage.

Levin, M. (1988, Winter). Caring new world: Feminism and science. *The American Scholar, 57,* 100–106.

Lewis, G. L. (1986). *Career interruptions and gender differences in salaries of scientists and engineers.* (Working paper for the Office of Scientific and Engineering Personnel, National Research Council). Washington, DC: National Academy Press.

Lipman-Blumen, J. (1972). How ideology shapes women's lives. *Scientific American, 226*(1), 34–42.

Long, J. S. (1978). Productivity and academic position in the scientific career. *American Sociological Review, 43,* 889–908.

Long, J. S. (1990). The origins of sex differences in science. *Social Forces, 68,* 1297–1315.

Long, J. S. (1992). Measures of sex differences in scientific productivity. *Social Forces, 71,* 159–178.

Long, J. S., Allison, P. D., & McGinnis, R. (1993). Rank advancement in academic careers: Sex differences and the effects of productivity. *American Sociological Review, 58,* 703–722.

Long, J. S., & McGinnis, R. (1981). Organizational context and scientific productivity. *American Sociological Review, 46,* 422–442.

Lotka, A. J. (1926). The frequency distribution of scientific productivity. *Journal of the Washington Academy of Sciences, 16*(12), 317–323.

Maccoby, E. E., & Jacklin, C. N. (1974). *The psychology of sex differences.* Stanford: Stanford University Press.

Matyas, M. L. (1991). Women, minorities and persons with physical disabilities in science and engineering: Contributing factors and study methodology. In M. L. Matyas & S. M. Malcom (Eds.), *Investing in human potential: Science and engineering at the crossroads* (pp. 13–36). Washington, DC: American Association for the Advancement of Science.

McBay, S. M. (1987, November 30). Science for women and minorities. *The Scientist, 1*(26), 29.

Merton, R. K. (1973). *The sociology of science: Theoretical and empirical investigations.* Chicago: University of Chicago Press.

Moen, P. (1988). *Women as a human resource* (Manuscript NSB/EHR 89-05). Washington, DC: Sociology Program, Division of Social and Economic Science, National Science Foundation.

National Academy of Sciences (NAS). (1979). *Climbing the academic ladder: Doctoral women scientists in academe.* (Report of the committee on the education and employment of women in science and engineering, Commission on Human Resources). Washington, DC: National Academy of Sciences.

National Research Council (NRC). (1981). *Postdoctoral Appointments and Disappointments.* (Report of the committee on a study of postdoctorals in science and engineering in the United States, Commission on Human Resources). Washington, DC: National Academy Press.

National Research Council (NRC). (1983). *Climbing the ladder: An update on the status of doctoral women scientists and engineers.* (Committee on the education and employment of women in science and engineering, Office of Scientific and Engineering Personnel). Washington, DC: National Academy Press.

National Science Foundation (NSF). (1986). *Women and minorities in science and engineering* (NSF 86–301). Washington, DC: National Science Foundation.

National Science Foundation (NSF). (1988). *Science and engineering doctorates: 1960–86* (NSF 88–309). Washington, DC: National Science Foundation.

National Science Foundation (NSF). (1990). *Women and minorities in science and engineering.* (NSF 90–301). Washington, DC: National Science Foundation.

Nowotny, H. (1991). Mixed feelings: Women interacting with the institution of science. In J. R. Blau & N. Goodman (Eds.), *Social roles and social institutions: Essays in honor of Rose Laub Coser* (pp. 149–165). Boulder, CO: Westview.

O'Leary, V. E. (1988). Women's relationships with women in the workplace. In B. A. Gutek, A. H. Stromberg, & L. Larwood (Eds.), *Women and work* (Vol. 3), (pp. 189–213). Newbury Park, CA: Sage.

Price, D.J.D.S. (1963). *Little science, big science.* New York: Columbia University Press.

Reskin, B. (1976). Sex differences in status attainment in science: The case of the postdoctoral fellowship. *American Sociological Review, 41,* 597–612.

Reskin, B. F. (1978). Sex differentiation and the social organization of science. In J. Gaston (Ed.), *Sociology of science* (pp. 6–37). San Francisco: Jossey-Bass.

Reskin, B. F. (1994). Sex segregation: Explaining stability and change in the sex composition of work. In P. Beckmann & G. Engelbrech (Eds.). *Arbeitsmarkt für Frauen 2000—Ein Schritt vor oder ein Schritt zurück?* (pp. 97–115). Nürnberg, Germany: Bundesanstalt für Arbeit.

Retherford, R. D., & Sewell, W .H. (1991). Birth order and intelligence: Further tests of the confluence model. *American Sociological Review, 56,* 141–158.

Roe, A. (1952). *The making of a scientist.* New York: Dodd, Mead.

Roos, P. A. (1985). *Gender and work: A comparative analysis of industrial societies.* Albany: State University of New York Press.

Rosenfeld, R. A. (1984). Academic career mobility for women and men psychologists. In V. B. Haas & C. C. Perrucci (Eds.), *Women in scientific and engineering professions* (pp. 89–127). Ann Arbor: University of Michigan Press.

Rossi, A. S. (1977). A biosocial perspective on parenting. *Daedalus, 106,* 1–31.

Sabljic, A., & Trinajstic, N. (1988). A formula for rating scientists. *Periodicum Biologorum, 90*(3), 397–399.

Salk, J. E. (1989, August 9–13). *How did I get there? Agents, events, and kin in the mobility accounts of elite young business professionals.* Paper presented at the American Sociological Association Annual Meeting, San Francisco.

Schooler, C. (1972). Birth order effects: Not here, not now! *Psychological Bulletin, 78,* 161–175.

Simmel, G. (1950). *The sociology of Georg Simmel* (K. H. Wolff, Trans. & Ed.). New York: Free Press.

Simonton, D. K. (1988). *Scientific genius: A psychology of science.* Cambridge, UK: Cambridge University Press.

Sonnert, G. (1991, November 15–17). *What makes a good scientist? Determinants of peer evaluation among biologists.* Presented at Society for Social Studies of Science Annual Meeting, Cambridge, MA.

Steelman, L. C. (1985). A tale of two variables: A review of the intellectual consequences of sibship size and birth order. *Review of Educational Research, 55,* 353–386.

Stein, A. H., & Bailey, M. M. (1973). The socialization of achievement orientation in females. *Psychological Bulletin, 80,* 345–366.

Sulloway, F. J. (1989, February 3). *Orthodoxy and innovation in science: A multivariate synthesis.* Talk given at the Harvard Department of the History of Science.

Tannen, D. (1990). *You just don't understand: Women and men in conversation.* New York: Ballantine.

Vetter, B. (1984). Changing patterns of recruitment and employment. In V. B. Haas & C. C. Perrucci (Eds.), *Women in scientific and engineering professions* (pp. 59–74). Ann Arbor: University of Michigan Press.

Vetter, B. (1987). Women's progress. *Mosaic, 18*(1), 2–9.

Visher, S. S. (1947). *Scientists starred 1903–1943 in "American men of science."* Baltimore: Johns Hopkins Press.

Walster, E., Cleary, T. A., & Clifford, M. M. (1971). The effect of race and sex on college admissions. *Sociology of Education, 44,* 237–244.

Weiner, B. (1974). Achievement motivation as conceptualized by an attribution theorist. In B. Weiner (Ed.), *Achievement motivation and attribution theory* (pp. 3–48). Morristown, NJ: General Learning Press.

Weitzman, L. J. (1984). Sex-role socialization: A focus on women. In J. Freeman (Ed.), *Women: A feminist perspective* (3rd ed.), (pp. 157–237). Palo Alto, CA: Mayfield.

Widnall, S. E. (1988). Voices from the pipeline (AAAS presidential lecture). *Science, 241,* 1740–1745.

Wright, R., & Jacobs, J. A. (1994). Male flight from computer work: A new look at occupational resegregation and ghettoization. *American Sociological Review, 59,* 511–536.

Zajonc, R. B. (1976). Family configuration and intelligence. *Science, 192,* 227–236.

Zuckerman, H. (1987). Persistence and change in the careers of men and women scientists and engineers: A review of current research. In L. S. Dix (Ed.), *Women: Their underrepresentation and career differentials in science and engineering* (pp. 123–156). Washington, DC: National Academy Press.

Zuckerman, H. (1988). The sociology of science. In N. J. Smelser (Ed.), *Handbook of sociology* (pp. 511–574). Newbury Park, CA: Sage.

Zuckerman, H. (1989). Accumulation of advantage and disadvantage: The theory and its intellectual biography. In C. Mongardini & S. Tabboni (Eds.), *L'opera di R. K. Merton e la sociologia contemporanea* (pp. 153–176). Genoa: Edizioni Culturali Internationali Genova.

Zuckerman, H., Cole, J. R., & Bruer, J. T. (Eds.). (1991). *The outer circle: Women in the scientific community.* New York: Norton.

INDEX

administrative duties, 138–139
admission rules, 10
advantages, accumulation of, x, xvi,
5, 21, 22, 24, 25, 27, 63, 66, 110–122,
164
advisors: collaboration with, 84 table,
85, 92; effect on career, 117; gender,
85, 86, 97, 99 table, 100, 102 table,
109 table, 118–119; quality of, 110;
rank, 78, 79, 84 table, 88, 92, 104,
112
affirmative action, 125, 128
age: and graduate school interruptions, 80; and parental education
level, 69; and rank, 43, 44 table
aggressiveness, 14, 18

behavior, achievement-oriented, 13
brother-impact theory, 70, 71

capital, human, 65–122
career: achievement expectations, 13;
alternatives, 27; background variables, 67–76; decisions, xvi; deficit
model, 9–12, 19, 126, 132, 140, 144,
164; difference model, 12–17, 19,
126, 132, 140, 164; effect of childhood losses on, 72–74; and family
status, 114–116; fellowship as
stage in, 28–30; gender disparities,
xiii, 9–17 *passim*; graduate school
effect on, 82–83; kick-reaction
model, xi, 16, 17, 154–155; and

marriage, 115–116, 156–162; obstacles, xiv, 10, 16, 17, 18, 19, 30, 51,
124, 126, 127–133, 156–162; parental effect on, 75; and parenthood,
114–115; predicting outcome, xiv;
sibling effect on, 69–72, 74–75;
stages, 65–122; and type of fellowship, 51–52
collaboration, 116–119, 133–137, 141–
142; with advisor, 84 table, 85, 92;
avoidance of, 136–137; benefits of,
106–107; family-style, 141–142;
during fellowship stage, 95; research, 88, 98, 133–134, 135 table;
styles, 77, 78 table, 98, 101, 106,
112, 116–119; subservient, 117;
women's difficulties in, 18
communalism, 3
compensation model, 65, 79
competition, ix, 14
constructivism, 6, 23
critical mass, 11, 51

deficit model, 9–12, 19, 126, 132, 140,
144, 164, 165
difference model, 9, 12–17, 19, 126,
132, 140, 164, 165
disadvantages, accumulation of, x,
xvi, 5, 21, 22, 24, 25, 27, 63, 66, 110–
122, 164–165
discrimination, 20–23, 127–139;
avoidance of, 131; awareness of,
129; compliance with, 130–131;

183

gender differences in, 39–63; gender handicaps in, 26; internal factors, 13, 14; luck in, 13, 14, 125, 126; and time spent in graduate school, 79–80
surrogate-son theory, 70, 71

tenure, xvi, 45–47; discrimination, 18, 127; glass ceilings in, 50; marginalization in, 45, 46; predicting, 49; and rank, 45
thinking, women's ways of, 12
thresholds, x, 25, 26, 27, 39, 50, 51, 63, 163, 164
tokenism, 11, 51

universalism, 3, 4, 8
universities: denial of access to, 10; glass ceilings in, 50; prerequisites for careers in, 29; prestige values of, 47–48; social system of, 42

values: academic, 42; scientific, 42; social, 40

women: academic rank, 30–31, 31 table, 42–45; and administrative duties, 138–139; background variables in career choice, 65–76; in biological sciences, 35, 45, 46, 51, 57; career alternatives, 27; career obstacles, xiv, 10 *passim;* collaborative styles, 98, 101, 106, 116–119; connectedness among, 141; discouraged from science careers, 11, 14, 16; domestic responsibilities, 19, 114–115, 124, 156–162; and double standard, 132–133; exclusion of, 10, 18; expectations for, 13; in labor force, 12; lack of financial support, 18; marginalization of, 2, 15, 18, 45, 46, 141, 155, 164; occupational under-representation, 7, 43; parental effect on, 67–69, 68 table, 75; perfectionist approach, 148–149; professional conduct, 143–146; professional rank, 30–31, 31 table, 42–45; publication productivity, 52–63, 90, 97, 107, 113, 149–151; reasons for taking fellowships, 97; relations with advisor, 85, 108; research styles, 147–148; role socialization, 12–17; sibling effect on, 69–72, 75; and standards of science, 151–155; status in science, 30–32; as strangers, 1, 2, 9, 11, 136, 164; structural barriers against, 7, 8; as supplemental income-earners, 40; and tenure, 45–47; time spent in graduate school, 80–81; ways of thinking, 12

ABOUT THE AUTHORS

Gerhard Sonnert studied sociology, history, and geography in Germany and the United Kingdom. He received his doctorate in sociology from the University of Erlangen, Germany, in 1986, and a Master of Public Administration degree from Harvard in 1988. Since then he has worked at Harvard for Project Access, a large-scale research project on scientists' career patterns. **Gerald Holton** is Professor of Physics and Professor of the History of Science at Harvard University. His research into the scientific, social, and epistemological roots of the achievements and failures of major twentieth-century scientists led him to initiate this research project on the careers of scientists today.